INSECTS IN FLIGHT

INSECTS IN FLIGHT

A glimpse behind the scenes in biophysical research

by WERNER NACHTIGALL

Director of the Zoological Institute, University of Saarlandes,
Saarbrücken, West Germany

Translators:
Harold Oldroyd
Department of Entomology, British Museum (Natural History)
Roger H. Abbott
Department of Zoology, University of Oxford, England
Marguerite Biederman-Thorson
Max Planck Institute for Behavioural Physiology,
Seewiesen, West Germany

WITHDRAWN

McGraw-Hill Book Company

NEW YORK	DÜSSELDORF	NEW DELHI
ST. LOUIS	JOHANNESBURG	PANAMA
SAN FRANCISCO	KUALA LUMPUR	RIO DE JANEIRO
	LONDON	SINGAPORE
	MEXICO	SYDNEY
	MONTREAL	TORONTO

COLLEGE OF LAKE COUNTY

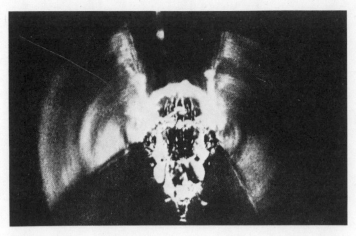

JACKET PICTURE

Here is a 'bluebottle', the blowfly Calliphora erythrocephala, *about twelve times natural size. This is not a flash picture, but an exposure of about one-fiftieth of a second, during which time the wing has beaten up and down four times, producing the effect of a glistening arc for each wing. If the two wings are carefully compared, it will be seen that the one on the right of the picture (the fly's left wing) is beating through a distinctly larger arc than the other, and that it is travelling further upwards as well as further downwards. It will also be noticed that the fly has turned its head slightly to its right (our left). The fly looks as if it wanted to turn right, and the stronger beat of the left wing would certainly have had this effect if the fly had not been firmly stuck to its support by its abdomen. Eighteen high-intensity, low-voltage lamps were needed to take this difficult picture, and the set-up is shown in Plate 8.*

The fact that the tethered fly is still performing the natural motions of flight is shown by the way in which the legs are held, in the attitude typical of normal flight. The hind legs are stretched out backwards, and in the picture they merge into the upper edges of the arcs of the two wings. The middle legs are held obliquely forwards and the fore legs are stretched out straight forwards. The tarsi, or 'feet', of these two pairs of legs are visible as light brown blobs in front of the head. It is interesting that the long, backwards-facing hairs on the thorax and abdomen have taken up a position exactly in line with the flow of air over the surface of the body.

This translation © George Allen & Unwin Ltd 1974

© 1968 by Heinz Mooz Verlagsgesellschaft München, Gräfelfing vor München

Filmset in Great Britain by Photoprint Plates Ltd, Rayleigh, Essex, in 10 on 13 point Bembo
Printed by Lowe & Brydone (Printers) Ltd, Thetford, Norfolk.

Library of Congress Cataloging in Publication Data

Nachtigall, Werner.
 Insects in flight.

 Translation of Gläserne Schwingen.
 Bibliography: p.
 1. Flight. 2. Wings. 3. Insects—Anatomy. 4. Animal mechanics.
I. Title.
QL496.7.N313   595.7'01'852   70–172030
ISBN 0–07–045736–0

FOREWORD

The restless times in which we live encourage superficiality, and we are all inclined to take too much for granted. A surfeit of scientific marvels has dulled our sense of curiosity, and we are losing the urge to find out things for ourselves. Today we stand more than ever in need of a spirit of inquiry, and a renewed capacity for wonder. The study of nature can help to restore these to us, because the world of nature abounds in fantastic things which the science of biology interprets for us by showing how the structure of living organisms is closely related to their function.

It is my belief that such studies are not out of touch with the needs of everyday life, and that the scientist is not taking refuge in an ivory tower whatever line of research he chooses to pursue, as long as he is making new discoveries and uncovering facts previously not known. Ever since I was in high school I have had a rash idea that I might write a book of popular science somewhere on the boundary between biology and physics, say about the biophysics of the flight of insects. I am actively employed in this field of study at the present time, and so can write about it with some personal experience.

I want to try to convey to the reader something of the joy of discovery that comes to a research worker, as well as the trials and problems that are an inseparable part of research work, and I should like to communicate a little of my own sense of wonder at the sheer inventiveness of nature. I have been at pains not just to give the reader a catalogue of recent discoveries in biophysics, but to try and make him feel that he is participating in the laborious and long-drawn-out process of conducting experiments; to give him a peep behind the scenes into the workaday world of the research worker.

Although explanations are given in a simple way that an interested layman should be able to follow, I have been particularly careful to make sure that the essentials of the matter are correct, and that all the details are as accurate as possible. In this way I hope to have written a book that both the general reader and the student may read with profit, and perhaps even other research workers may derive pleasure from it.

CONTENTS

Foreword   *page*

1. The external and internal structure of an insect — 9
2. The importance of classification — 12
3. The form and construction of the wings of insects: glistening scales and glassy membranes — 13
4. Sail butterflies gliding on the seashore — 21
5. Gliding and soaring are flying without muscle power — 21
6. What are forces and how are they represented? — 23
7. A little about the aerodynamics of the insect wing — 24
8. Painstaking experiments mean safe aircraft — 26
9. In the realm of thousandths of a second — 30
10. Insect flight in the wind tunnel — 35
11. 'What a lot of money to spend on a tiny fly!' — 37
12. Two years of preparation—two seconds of photography — 38
13. The fascinating play of the wing movements — 41
14. One two-hundredth of a second in the life of a bluebottle — 46
15. How the wing beats generate aerodynamic forces — 47
16. Glittering wings flit through the room — 49
17. Watching insects for pleasure — 55
18. The precision mechanism that drives the wings — 59
19. The wing joint as a tiny click mechanism — 65
20. The mighty flight motor, its performance, and the problem of supplying it with oxygen — 68
21. Fuels for insect flight — 78
22. Cooling the engine — 80
23. The wonderful substance resilin — 82
24. The construction of the thorax, light yet strong — 83
25. The deer botfly is not a living projectile, and it does not fly at supersonic speed — 87
26. How do pygmy insects fly? — 89
27. Hoverflies, the helicopters of the animal kingdom — 92
28. What controls the muscle twitches? Or: $30,000 worth of electronics — 93
29. The grasshopper beats its wings twenty times per second; the gnat, a thousand times: there are slow and fast flight muscles — 95
30. The fine structure of muscle as seen under the electron microscope: the smallest chemical factories and the mechanism of muscle contraction — 105
31. The accelerator pedal for the flight motor: nerve impulses regulate the delivery of muscle energy — 108
32. The steering gear: how does an insect fly right and left? — 110
33. Insects walk on six legs — 116
34. A catapult launch is nothing remarkable for the fly — 117
35. How does an insect land upside down on the ceiling? — 119
36. What does an insect do with its wings when it is not flying? — 120
37. Migratory flight of locusts and butterflies — 122
38. The migrant habits of our native insects — 129
39. How foraging bees find their way back to the hive: an exercise in direction-finding and aerial navigation — 131
40. How is flight velocity regulated? — 134

page

41. Sense organs are the insect's flight control instruments   139

42. By their song ye shall know them   141

43. Aerial combat at night, with supersonic direction-finding and 'sonar jamming'   142

44. Comparison between insect and human flight   144

Bibliography   147

Illustrations acknowledgements   150

Index   151

A bluebottle irritates us with its buzzing and we squash it. What have we destroyed?

It was an insect, with head, thorax and abdomen, with two wings and six legs. It was an invertebrate animal, that is, it had no internal skeleton of bone; instead it had an external skeleton in the form of an armoured shell. When flying at its fastest it could cover 250 lengths of its own body per second, beating its wings 300 times while doing so. We have destroyed a masterpiece of intricate construction, built up from millions of cells, grouped into organs of several dozen different types, yet which nevertheless all worked together in perfect harmony. Fig. 1 gives an outline of the structure of a simple, primitive insect, less highly developed than our blowfly.

The head of the fly bears two impressive-looking compound eyes, each a red-brown hemisphere consisting of thousands of hexagonal facets pressed closely together and forming a complex of light-sensitive cells. The components of the insect's brain, called the first, second and third optical ganglia and which lie in line one behind the other, interpret the numerous light impulses. Three simple eyes, called ocelli, stand on the vertex, or top, of the head. The underside of the head bears a proboscis which is capable of being greatly extended, and its tip is equipped with a sucking sponge, through which the fly can draw up liquid food through several thousand minute tubules. The proboscis also has taste cells and chemoreceptive organs with which the quality of the food can be assessed. Massive salivary glands open into the mouth cavity and release saliva which begins to digest the food even before it has reached the intestine. Small feelers called palpi help to guide the proboscis and direct it towards the food material.

In between the compound eyes stand the two antennae, which have a remarkable structure. Their function is to help the fly to orientate itself, by vibrating in response to the flow of air over the fly's head, and also to regulate the speed of flight. The brain consists of a fusion of ganglia of nerve-complexes as we have seen, and its function is to receive and co-ordinate the impulses sent in by the many sense organs. In turn the brain sends out commands to the various muscles of the fly's body. A system of horny rods and flanges inside the head keeps the various organs in place and also provides attachments for the muscles, especially for the powerful muscles necessary to operate the proboscis.

The digestive system begins in the head with the salivary duct. The entire head is mounted on horny knobs which

project from the front surface of the second section of the body, the chest or thorax. Everything that passes between head and thorax must go through the occipital foramen, a small hole at the back of the head: nerves, air ducts of the

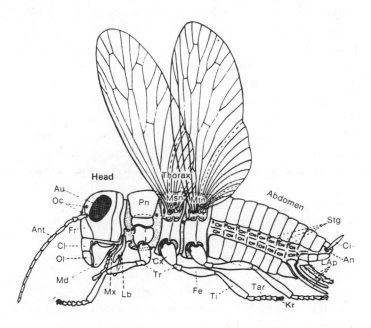

Fig. 1. Semidiagrammatic drawing of a four-winged insect of the locust type in side view (above) and in longitudinal section (below). Upper figure: An, anus; Ant, antennae; Au, compound eye; Ci, cerci (anal lamellae) Cl. clypeus: Cx, coxa; Fe, femur; Fr, frons; Kr, claw; LAp, egg-laying apparatus (ovipositor); Lb, labium (lower lip); Md, mandible (upper jaw); Msn, mesonotum; Mtn, metanotum; Mx, maxilla (lower jaw); Oc, ocelli; Ol, labrum (upper lip); Pn, pronotum; Stg, spiracle (external opening of respiratory system); Tar, tarsus; Ti, tibia; Tr, trochanter. Lower figure: An, anus; Ao, aorta; DAm, dorsal ampulla; DDi, dorsal diaphragm; G, brain; Gp, gonapophyses; H, heart; HD, hind gut; KM, proventriculus; Kr, crop; Lb, labium; MD, midgut; Md, mandible; Mo, mouth opening; N, nerve cord; Oe, oesophagus; Ov, ovary; PM, peritrophic membrane; R, rectum; So, supraoesophageal ganglion; Sp, salivary gland; VDi, ventral diaphragm.

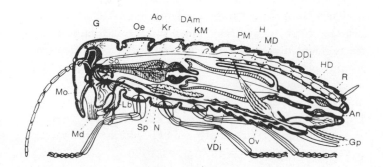

tracheal system, blood and food. A great many tiny muscles emerge from the corresponding hole at the front of the thorax and are attached to the back surface of the head, or occiput. By nicely co-ordinated contractions of these muscles an insect can turn its head to direct the various sense organs as well as the proboscis, in any desired direction, and to hold the head in that position. This articulation works very much like a ball-and-socket joint, except that its range of movement is more restricted, though some insects have exceptional powers of turning the head.

The cavity of the thorax is almost entirely filled with the muscles of flight, which form an astonishingly large bundle. Hardly any space is wasted and the individual fibres seem to be in a disorderly arrangement, lying alongside, across and between one another. They arise from the thoracic wall, from strong horny ledges sticking out from this, or from the coxae, or hipjoints of the legs. The muscles run to another point on the thorax, or else directly to the axillary sclerites, small pieces of hard chitin upon which the wings articulate.

If a fly is dissected, a total of twenty-one bundles of flight muscles will be found to lie in each lateral half of the thorax! The function of these muscles is to generate thrust by the wings against the surrounding air, and also to steer the fly in flight. The smallest muscles are only a few hundredths of a millimetre thick. Massive trunks of the tracheal system of breathing tubes can be seen amongst the musculature, sending out branches into all the bundles of muscle, often in a regular sequence. These air tubes supply the extraordinarily high demand for oxygen that comes from the flight muscles.

The tracheal system communicates with the outside air through two pairs of large openings called spiracles, one pair near the front of each thorax side, and the other further back. Each spiracle has a fringe of bristly hairs which exclude dust. The intestine and the longitudinal nerve cord of the fly lie below the masses of muscle, isolated in a kind of tunnel formed from inturned ledges of the thoracic wall. All six legs are attached to the underside of the thorax by ball-and-socket joints. At their tips (tarsi) they are equipped with claws and adhesive pads (pulvilli) which enable the fly to walk about on smooth vertical surfaces such as windows. The tarsi also bear organs of touch and taste.

During the embryonic development of a fly, the thorax is clearly composed of three segments, but in the adult these become so thoroughly fused into a single tough capsule that the segmentation is no longer obvious, though it is indicated by the location of the three pairs of legs. The fore legs are attached to the prothorax, the middle legs to the mesothorax, and the hind legs to the metathorax. In true flies (order Diptera) only the middle segment, or mesothorax, bears a pair of wings. Interestingly enough, during embryonic development the hind segment, or metathorax, also bears a

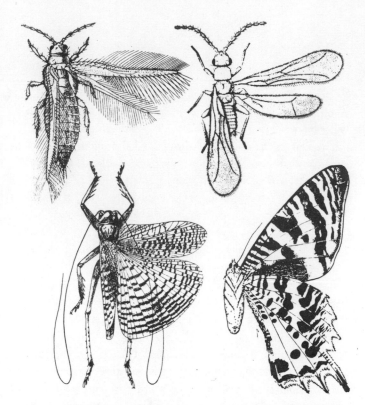

Fig. 2. *The four basic types of functional insect wing, illustrated by one species from each of four different orders. In each insect the wings on the right side are shown as if set, while those on the left side are either shown folded or omitted. Left upper: stiff, fast-moving, bristly wings of the thrips* Taeniothrips inconsequens *(Thysanoptera). Right upper: stiff, fast-moving, membranous wing of* Zorotypus brasiliensis *(Zoraptera) (after Sylvester, 1913); winged female. Left lower: folding, slow-moving, membranous wing of a bush-cricket,* Pardogryllacris spuria *(Orthoptera). Right lower: stiff, slow-moving, scaly wing of a moth,* Urania ripheus *(Lepidoptera).*

pair of wing buds, but these soon begin to develop differently, and become the halteres, or balancers. They have the form of miniature drumsticks and vibrate rapidly up and down during flight. They have been shown to act as stabilisers, correcting yawing and other deviations from the correct flight attitude, but it is probable that an even more important function is to stimulate the body ready for flight (see Chapter 41).

Parts of the meso- and metathorax combine to provide the two complex articulations upon which the wing moves. Like the head, the thorax narrows down posteriorly to a small aperture, through which all the supply systems must pass to reach the third and last section of the body, the abdomen. The abdomen is narrow at its base, but quickly swells into an elongate capsule which gradually narrows again to a blunt tip. This capacious structure is built up from a series of horny rings which fit into each other like a telescope, and are joined together and sealed by membranes. This form of joint, which also occurs between the head and thorax and thorax and abdomen, accounts for the name 'insect', which means 'cut into'.

The three main divisions of the body, head, thorax and abdomen, each have a different part to play in the life of the insect. We have already seen that the head, with its sense organs and proboscis, is almost solely preoccupied with orientation and feeding. The thorax, with its massive musculature, with its wings and legs and with its specialised systems for stabilisation, is concerned exclusively with loco-motion. What functions are left for the abdomen to perform?

The chief functions remaining are those of nutrition and reproduction, and the organs of digestion, absorption and excretion, and the genital systems do in fact lie in the abdomen. Here lie the intestine, the stomach, the voluminous digestive glands, the excretory organs (Malpighian tubules), and finally the hind gut. The last extends back to the anus, and disposes of waste matter by expelling it as semiliquid faeces, or frass. The part of the usable food that is surplus to im-mediate requirements is stored as globules of fat in the fat body, an amorphous mass which fills all the vacant spaces between the organs of the abdomen.

The abdomen also accommodates the sex glands, in male flies these are the testes and in females the ovaries. Externally the male has a penis and the female an ovipositor. In female flies related to the bluebottle the ovipositor is a telescopic tube which can be extended a surprising distance when eggs are being extruded. Sometimes the ovaries and the ripe eggs lying in the oviducts take up so much space that the abdomen is full to bursting, with the horny rings pushed wide apart and the intervening membrane stretched until it almost tears. At these times the abdomen takes on a whitish yellow colour since the mass of eggs is visible through the stretched membranes. This condition is conspicuous even to the naked eye, and if the fly is seized and gently squeezed the ovipositor will extend itself.

The nerve system of an insect runs down the middle of the belly with branches extending into the tip of the abdomen, and this arrangement can be seen very clearly in flies. A series of nerve complexes, or ganglia, in the abdomen and thorax control various local secondary functions, but are always under the overall control of the brain.

The heart lies above, or dorsally, to the other organs, and extends into the abdomen, though the non-biologist would instinctively look for the heart in the chest or thorax. It is an elongate tube which is set into rhythmical contraction by segmentally placed muscles. The blood flows into the heart from the general body cavity by means of pores called ostia, and the contractions drive it forwards as in an aorta. There are a number of openings through which the blood can escape again into the body cavity, and circulate among the various organs. Occasionally partitions, or septa, occur which divert the blood into a particular part of the body, for example in the legs. Furthermore, the blood of insects does not (except very rarely) contain any red blood corpuscles like those found in human blood. As everyone knows, red corpuscles exist to transport oxygen, and without them insects have evolved a totally different method of respiration. This is the system of air tubes, or tracheae, already mentioned.

Now we can understand how insects can manage with such a slow circulation of blood through such an inefficient circulatory system, full of holes. Oxygen is carried from the atmosphere direct to each muscle fibre, and almost to every individual cell. The heart does not have to pump blood through the capillary network of a closed circulatory system, nor is there urgent need for the blood to circulate quickly. Because the blood does not take up oxygen there is no need for insects to have lungs like air-breathing vertebrate animals. Indeed, this is one of the most significant differences between the internal organisation of an insect and a vertebrate. The chief functions of an insect's blood are to transport dissolved food materials, and to maintain a correct balance of dissolved salts in the internal fluids.

Out of a single cell, the fertilised ovum, arose the whole of this complicated machine that we call a 'bluebottle', a small, blue-black, troublesome buzzing creature. Looking at things from our human standpoint we often commit a serious error of observation. We assume that small creatures must be of simpler construction than big ones, because nature 'didn't have so many bits to make'. This is wrong. We humans are usually a shade less than two metres tall, and of exceedingly complex structure. We have everything that we need for our way of life. The little fly is quite differently constructed, but—except in the matter of brain development —just about as complex as a man. Moreover, the fly too has everything necessary for its way of life. Yet a single stroke kills a fly, destroying completely this diminutive, incredibly complex, wonderful nuisance.

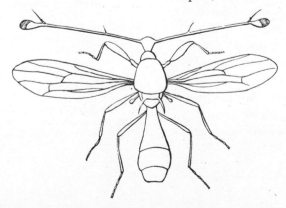

2. The importance of classification

Taxonomy—systematics, or classification—can often be a tedious subject, but we need to take a quick look at the subdivisions into which the animal kingdom is classified. It is true that our concern is only with winged insects, but we ought to be aware of how our insects fit into the general classification of living things. The living world is divided into two kingdoms, animals and plants, and each of these is subdivided into *phyla*. The animals with segmented legs, or Arthropoda are a phylum: examples of others are algae, flowering plants, Echinoderms (sea-urchins), and of course vertebrate animals. Each phylum is further subdivided into a series of *classes*: the most important classes of Arthropoda are Crustacea, Myriapoda (centipedes and millipedes), spiders, and finally, Hexapoda, or insects.

Insects are thus a class of animals, and like other classes they are subdivided into a number of orders. There are twenty-four orders of insects—though different authorities recognise a slightly different number—and the great majority of insects have wings. In Table 1 the scientific names printed in italics are those of orders of insects which include at least some winged members.

Even a quick glance at this table shows immediately that the possession of wings is widely distributed through the class Insecta, and is common to insects of strikingly different appearance, and with wings of very different construction. The general plan of the wings, their fine detail, and even their operating mechanism and means of propulsion, can take a wide variety of forms. Fleas have no wings, and so are unable to fly out of the pelts of dogs or rats, or from beneath human clothing. These insects have more use for a highly developed power of jumping than for flight. Lice, too, including the feather lice of birds, would have little use for wings. These orders of parasitic insects have more use for powerful claws on the tarsi, or feet, with which they can haul themselves about in the world of fur or feather in which they live. Termites (Isoptera) and true ants—which belong to the order Hymenoptera—need wings only for the mating flight, after which the wings break off. Below ground, and in the tunnels inside their nests, the wings would not be able to function and would soon become damaged.

A simple and easily understandable connection can be seen between the possession or not of wings and the way of life of the insect. Those which live predominantly in the open air have wings, while those which pass their lives in crevices or narrow passages, among hairs or feathers, are usually wingless. Insects of nearly all orders have at least rudimentary wing buds in the embryo, but in the groups just mentioned these rudiments never develop into functional wings. There are three exceptional orders of the most primitive insects, springtails, silverfish and the like, which do not have even embryonic wing buds. These are the primitively wingless insects, or Apterygota, and the other twenty-one orders are grouped together as the winged insects, or subclass Pterygota.

To summarise briefly the rest of the classification, each *order* is divided into *families*: for example the order Lepidoptera, the butterflies and moths, includes the families Papilionidae, Pieridae, Sphingidae, Noctuidae and very many others. Each family comprises a number of *genera*, and finally each genus is divided into *species*.

The scientific name of an animal always consists of two words, a *generic name* together with a *specific name*: for example the cabbage white butterfly is *Pieris brassicae*, *Pieris* being the generic name and *brassicae* the specific name. The cabbage white belongs to the family Pieridae, order Lepidoptera, class Insecta, phylum Arthropoda—working back upwards through the classification. The author of this book is no taxonomist, but even he is pained when he reads some of the garbled scientific names used by journalists.

Table 1 Summary of the classification of insects, according to von Lengerken (1953). Orders with winged members are shown in *italics*.

Class: INSECTA (HEXAPODA)
I Subclass: Apterygota. Primitive insects
  Order 1. Protura, semi-insects
        2. Collembola, springtails
        3. Thysanura, bristletails

II Subclass: Pterygota. Higher, or winged insects
  Order 1. *Ephemeroptera*, Mayflies
        2. *Plecoptera*, stoneflies
        3. *Odonata*, dragonflies and damselflies
        4. *Embioptera*, embiids, web-spinners
        5. *Orthoptera*, grasshoppers, locusts, crickets
        6. *Dermaptera*, earwigs
        7. *Isoptera*, termites
        8. *Copeognatha (Psocoptera)*, booklice
        9. Mallophaga Chewing lice, feather lice
      10. Anoplura, sucking lice
      11. *Thysanoptera*, thrips
      12. *Rhynchota (Hemiptera)*, bugs
      13. *Hymenoptera*, bees, wasps, ants, etc.
      14. *Coleoptera*, beetles
      15. *Strepsiptera*, stylopids
      16. *Neuroptera*, net-winged insects
      17. *Mecoptera*, scorpionflies
      18. *Trichoptera*, caddisflies
      19. *Lepidoptera*, butterflies and moths
      20. *Diptera*, two-winged, or true flies
      21. Aphaniptera, fleas

3. The form and construction of the wings of insects: glistening scales and glassy membranes

Fig. 2 shows the wingshapes of typical species from various orders of insects. And what a varied selection this is! Immediately one gets the instinctive feeling that the broad, ellipsoidal, rigid surfaces of a butterfly's wings must have quite different flight characteristics from the wings of a thrips, with their tattered fringes of hairs.

The wings of different insects also vary enormously in length. In Fig. 3 the insects are all drawn to roughly the same final *size*, but the reader must not be misled by this, and should realise that the *degree of enlargement* may be very different. Some fossil dragonflies found in deposits of the Carboniferous period had a wing length of over 50 centimetres, whereas that of some tiny Hymenoptera is less than 0.5 millimetre, a ratio of 1000:1. Suppose we compare this with a familiar example, a swan and a sparrow. Here the size ratio is only 10:1. As we know, the sparrow flies with a quick whirr of its wings, about fifteen strokes per second, whereas the swan flies in much more leisurely fashion, 'rowing' itself through the air at about one and a half strokes per second. We do not need much scientific knowledge to deduce from everyday observation that if two birds have wings of widely different size they will fly in a quite different manner, and with different rates of wing beat. If this is obvious in birds, where the size ratio is only 10:1, how much more must it be true of insects, where the size ratio is a hundred times greater.

This argument is not without its flaws. In particular, we have taken extreme examples of insects, but have only looked at everyday examples of birds. Suppose we take some more exceptional birds. An albatross may have a wing length of 2.5 metres, and a tiny hummingbird one of only 2.5 centimetres, giving a ratio of 100:1, reducing the gap somewhat. A comparison of wing beats reduces the difference still further. The humming bird can hover in one place, with its wings beating 50 times a second, whereas the albatross normally flaps its wings less often than once a second, and most of the time it soars with its wings motionless.

In this last example a conspicuous difference in wing shape supports the argument: the hummingbird, with broad, symmetrical thrusting wings; the albatross with elongate, gliding planes, arched in cross-section. So a few very simple comparisons have established the fact that the size of the wings, their shape, their rate of wing beat, and the resulting style of flight, are all closely linked with each other. We shall go into this in more detail later, but first we must look more closely at the insect's wings.

However much the outline of the wing may vary, it always

arises from a narrow, basal stump of essentially similar construction. The tip of this stump forms the outer member of the hinge mechanism, and its exact shape is critical for the correct beating of the wing, but it is so minute that it cannot be shown in the drawings of Fig. 3 at their scale of magnification. The wing margin may be either smooth or fringed, and every intermediate state exists. The little honeysuckle burnet moth has feathery wings, and under a hand-lens its outstretched wings look magnificent.

As we shall see later, the flight of such an insect, though peculiar, is not very different from the way in which water-fleas skip through the water. The sail butterfly has a long, narrow 'tail' projecting from the posterior angle of each hind wing, and this is probably associated with the gliding flight of this butterfly, from which it gets its name. On the other hand, mosquitoes and midges, with the extraordinarily high wing beat frequency of several hundred per second, have smooth, pointed wings with a regular outline. In contrast again, earwigs (Fig. 4) have wings like an extended fan, and which can be opened and closed like a real fan. Many earwigs manage to maintain a steady flight with this seemingly fragile mechanism. If a sail butterfly, an emperor moth, or even a humble cabbage white is held in the hand, and looked at from in front, it will be seen that a butterfly's wing is not flat like a sheet of paper, but convex. Anyone who has ever assembled a model of an aircraft from a balsa wood kit will remember that the wings had to be curved. In a similar way, the wing of an airliner must have convexity, and not just thickness (which of course the membranous wing of an insect does not have). Almost every insect wing has some degree of curvature, which not only gives it better static stability in gliding and dynamic stability when beating, but also improves its aerodynamic properties, at least in the bigger insects, as we shall see presently.

In the embryonic insect the wing consists of two horny layers of epidermis, between which the wing veins come into existence. The whole thing is a kind of pouch or sac, divided into compartments and filled with liquid rather like a plastic shampoo sachet. In particular, the wing at this stage is comparatively bulky. In those insects which undergo a complete metamorphosis the wings take shape within the pupa. The internal fluid and a good deal of the epidermis are later reabsorbed, so that eventually the thin layers of epidermis come together internally, except where they are kept apart by the veins (Fig. 5).

Because of lack of space in the pupa, the wings are closely

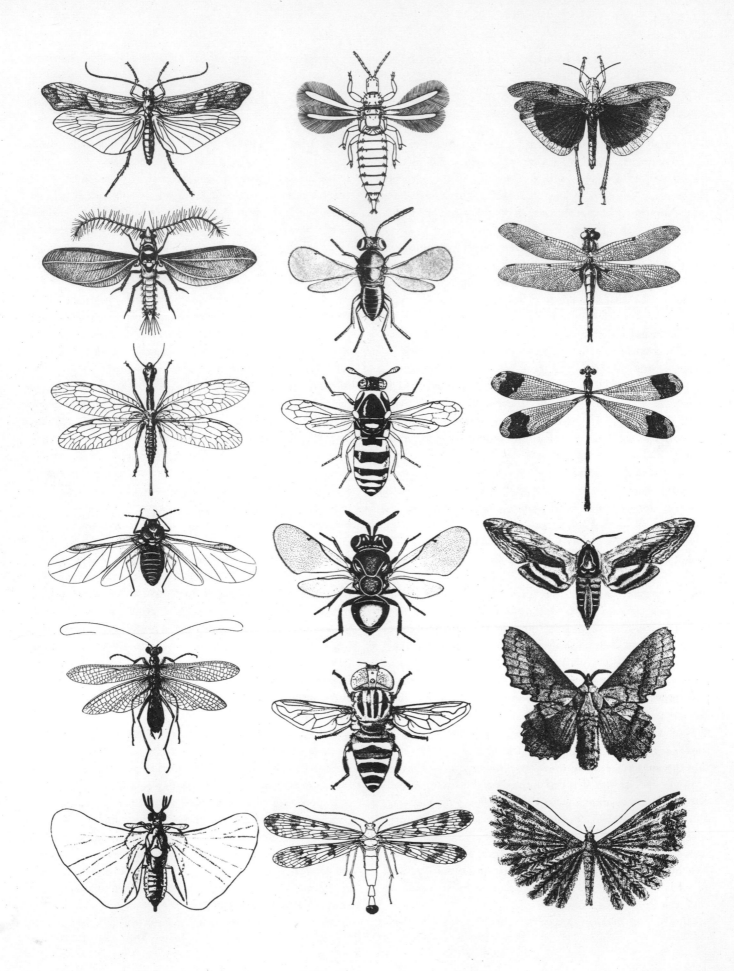

Left column from above downwards: *a caddis-fly* Limnophilus rhombicus; *a plant bug or coccid* Icerya purchasi, *male*; *a neuropteran, the snake-fly* Rhaphidia adanata; *a homopteran, the plant louse* Eriosoma lanigerum; *an orthopteran, the mantid* Mantoida brunneriana; *a strepsipteran,* Eoxeonos laboulbenei, *male*.
Centre column: *A thysanopteran, the thrips* Liothrips oleae; *a chalcid,* Coccophagus tschirchii, *male*; *a hymenopteran, the wasp* Celonites abbreviatus, *female*; *a hymenopteran, the parasitic wasp* Perilampus chrysopae, *female*; *a dipteran, the hoverfly* Lathrophthalmus quinquelineatus, *female*; *a mecopteran; the scorpionfly* Panorpa communis, *male*.
Right column: *an orthopteran, the meadow grasshopper* Dissosteira carolina; *a dragonfly,* Libellula quadrimaculata; *another dragonfly, the damselfly* Megalopropus coerulans; *a lepidopteran, the hawk moth* Hyloicus (Sphinx) ligustri; *another lepidopteran, the moth* Gastropacha quercifolia; *a lepidopteran, the plume moth* Orneodes cymatodactyla.

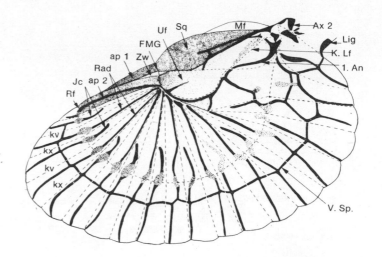

Fig. 4. *Left wing of the earwig* Forficula auricularia. *The wing is stretched out flat, and all the sclerotised structures are shown in black or shaded. Folds are drawn in dotted lines. The membrane is spread across a framework of radiating veins. alternately convex (kx) and concave (kv), which can open and close like a fan. When the fan has been closed in this way it is then folded once again, at right angles to the radiating veins, by means of the ring vein Rf, the distal section being brought over the rest. Where folding of the outer section like this bends each of the radiating (Rad) or intercalary veins (Jc) a short section is 'squashed', and these sections are shown on the plan as a dotted patch on each black vein. A further, downward folding of the wing takes place at the 'central hinge' (FMG), against the ulnar field (Uf), and the twice-folded wing is pressed against the rigid wing lobe, or squama (Sq). In Fig. 50 this folding process can be followed step by step. Other letters in Fig. 4 are: 1 An, first anal vein; ap 1 + ap 2, outer and inner apical field; Ax 2, second axillare; L.Lf, concave longitudinal fold in the ulnar field (this functions during flight as a locking bolt thereby making the wing doubly secure); Lig, wing ligament; Mf, marginal field; V.Sp., vena spuria; Zw, intermediate field.*

folded, even crumpled, and they expand and reach their full size only after the insect has emerged. This is brought about by the pumping of fluid into the intracellular spaces and a special secretion into the veins. The expansion of the wings is similar to that of the toys that are sold at fairs, where you blow into one end and a long paper tube unrolls. As soon as the wing veins are extended they fuse with the interconnecting membrane, and the wing has taken up its final form. The secretion inside the hollow veins hardens, and the veins change from hollow vessels filled with fluid into solid rods. From now on the shape of the wing is permanent. The short period of hardening-off is critical, and if anything goes wrong at this stage it cannot be corrected later. If by some mishap the wing cannot be fully expanded, and stays crumpled, it remains so, and is useless for flight.

The two membranes that have come together in the wing fuse, dry, and shrink to form a single membrane like a tough stretched skin. As soon as this has taken place the wings beat for the first time, and then the fully formed insect takes off and flies away. Anything from fifteen minutes to an hour is enough for the wing-hardening process in flies, butterflies, moths, and dragonflies. Plate 2 shows several stages of this phenomenon, but many more short descriptive stages would be needed to do full justice to this amazing process. For example, the final hardened membrane between the veins of the wing is at most one ten-thousandth of a millimetre thick!

The various veins do not lie in one plane, but some run at a higher level than the others. The membrane between veins therefore runs uphill in some places, and downhill in others, and so has a corrugated structure. This is particularly obvious when the wing is clear and glistening; something similar can be achieved by crumpling a sheet of tinfoil and then smooth-

ing it lightly on a table. It is roughly flat, but irregular ridges and furrows remain in it. If the sheet of tinfoil is smoothed, not on a flat surface, but round a tube, a curvature is superimposed on the pattern of ridges and furrows. This is the principle of the butterfly's wing.

The veins do not run at random, but according to a definite pattern, which is slightly different in each order of insects. Wherever the veins fork or cross each other they enclose a space known as 'cell'. Like the veins, the cells differ in different orders of insects, and each vein and each cell has its name. Although the wing venation is different in each member of a different order that we examine, a general pattern can be detected. Certain main veins and principal cells can be found in nearly all orders, and this is evidence that the wing venation of the different orders of insects is related, and that they are all evolved from a common ancestor.

In Table 1, p. 12, the twenty-four orders of insects are arranged in sequence, with the most primitive at the beginning, and the more highly evolved towards the end. Looking

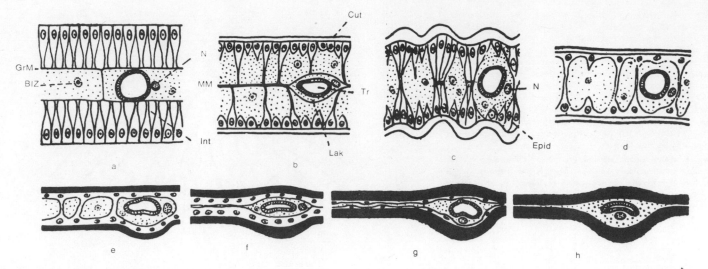

Fig. 5.   Successive stages of development of an insect's wing, shown in cross-section, from the original plan (a) to the finished wing (h). The embryonic plan (a) consists of two thick epidermal layers, each built up from elongate cells with intercellular cavities (Int) between them. Each epidermal layer rests on a basement membrane (GrM). These two basement membranes move towards each other and fuse together (b) into a central membrane (MM), while thick cuticular layers form on the two outer surfaces (Cut). After the final moult the wings are expanded by pumping blood into the intercellular spaces, while the epidermis cells (Epid) and the central membrane dwindle away (c, d, e, f). Most wings at this stage consist only of two glassy cuticular layers pressed against each other (g, h), with enough space here and there for a vein to pass between them. Other letters: BlZ, blood cells; Lac, lacunae; N, nerve; Tr, trachea.

at the wing patterns in Figs. 2 and 3 from this viewpoint, we can see that primitive insects have an elegantly simple wing venation. The main veins radiate fan-like from the base of the wing, with few cross-veins, and they seem all to be of much the same importance. The locust shows this type of venation very clearly. More highly evolved insects have more complicated wing venation. The individual veins run in quite different directions, particularly near the leading edge of the wing where the veins are thickened and much sub-divided, so that each section of the wing is different from the others. Only a few veins, and those towards the trailing edge of the wing, retain their 'primitive' pattern. This can be seen very well in the wings of flies, and of bees and wasps (Fig. 3, p. 14). It is immediately obvious that the primitive type of wing must be more pliable, and more easily twisted than the more advanced type; in the latter the stiffening of the leading edge means that only the posterior areas of the wing remain flexible.

It is not difficult to see that when the wings beat the primitive type will flex and deform quite differently from the more advanced type, and if this is so it is to be expected that the means by which the wing transmits power to the air will be entirely different in the two types. Both the static and dynamic forces acting on the wing will be different. There is thus an evident correlation between the level of evolution of an insect order and both the statics and dynamics of its wings. A few pages back we came to the conclusion that a relationship can be traced between wing length, wing shape, frequency of wing beat, and manner of flight of the insect.

This is how a scientist formulates a theory. He observes, compares, gathers experimental data, thinks over what he has accumulated and tries to fit it all together. Eventually he arrives at a possible explanation, which constitutes his hypothesis, though this is not yet conclusive and proves nothing. On the contrary, a hypothesis must be tested by an investigation specially planned so that it will show whether the hypothesis is true or false. Our present hypothesis might run as follows: certain details of wing venation occur together so frequently that we can assume that whenever they occur together in insects even of different orders they indicate an efficient flight mechanism.

Once we have established an association of this kind, and formulated a hypothesis, we can express it in algebraic symbols: 'if vein 1 in area a runs in direction x, vein 2 in area b in direction y, and vein 3 in area c in direction z, then the wing will be functionally efficient'. In other words, once the

Plate 1.   The left wing of the bluebottle fly Calliphora erythrocephala, enlarged about 25 times. Note the concentration of strong, rigid veins into the forward third of the wing, the reduction in both number and strength of the veins in the posterior third, and the very delicate hind margin composed of membrane only. At the base of the wing there is a smaller lobe, the alula, the function of which is still unknown. The leading edge of the wing is exceptionally strong, and is armed with bristle-like projections, which no doubt have their aerodynamic function, as must also the fine covering of hairs on the wing membrane.

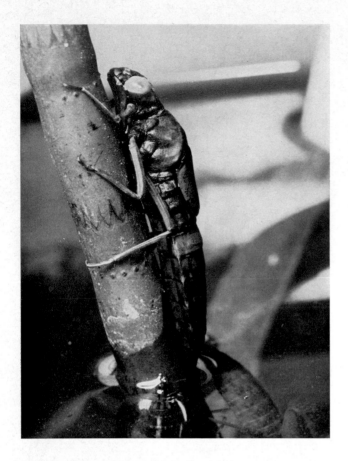 

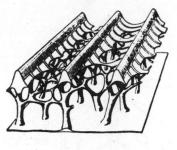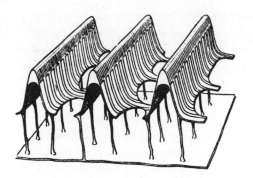

Fig. 6. *Structure of the scales on the wings of Lepidoptera.* Left to right: *a glandular scale from a butterfly showing the accessory disc at the base of the scale; fine structure of a clothing scale of the lattice type; fine structure of a scale of the corrugated type; a single scale of normal type.*

The fine structure of the scales was revealed by the electron microscope. The basal plate consists of a simple, relatively thick membrane, from which arise longitudinal ridges joined together by transverse bearers, and these again interlaced with cross members according to the type of scale, —lattice or corrugated—all these structures are bound together by a thin membrane cut away in places as shown in the drawing. The two parallel walls of the scale are held apart by a forest of stalks. A similar form of construction is used by man when he wishes to make a structure strong with the maximum economy of material, for example, corrugated paper, corrugated iron or asbestos, reinforced concrete, and all forms of sandwich materials such as plywoods and blockboards.

relationship has been investigated in sufficient detail for an association to be positively established and predictions made from it, we have taken the first step towards an understanding of insect flight. It must be pointed out that in practice we could not begin with a hypothesis like the preceding one because it goes straight to a general principle, and so is much too complex to form a basis for actual research. Working hypotheses normally confine themselves to a simple association between two small details. Research proceeds step by step, link after link is established, gap after gap is closed. Scientific research goes from hypothesis to theory, from theory to law. But let us return to our wings!

Higher and higher magnification always reveals more and more details. A wing does not consist just of veins and cells, but has many other and more minute structures. Hairs and scales may cover the membrane; the leading edge (costal margin) and the major veins may bear combs of bristles, and various kinds of spines or teeth, and the trailing edge (posterior margin) may have a regular fringe of hairs or scales. Everyone knows that Lepidoptera and caddisflies (Trichoptera) have wings covered with overlapping scales or hairs like the roof of a house. It is less widely known that a glassy transparent wing, such as many flies have for example,

may be coated with a regular array of many thousands of the minutest of hairs.

It seems fairly certain nowadays that all these wing embellishments must have their effects on its flying characteristics. A covering of scales, for instance, certainly increases the thrust of a wing, as well as having other secondary effects. These scales are responsible for most of the colour and pattern of butterflies and moths, and so play their part in mating and sexual selection, as well as in cryptic and warning coloration. It was not until the era of the electron microscope that the complex structure of a single scale could be appreciated. Fig. 6 shows some examples magnified from a few hundred to several thousand times, revealing a wealth of unsuspected detail, with structures like miniature roofs on stilts, with windows, mezzanine floors and toothed margins. A hundred thousand of these fine filigree structures may cover a single wing of a peacock butterfly.

An interesting feature of some wings is the existence of breaks in the veins and preformed lines of weakness. Where such a line of weakness crosses a vein there is a short, pale, colourless stretch. It is along such a line of weakness that ants break off their wings after taking part in the mating flight, and beetles fold their hind wings. Fig. 4 (p. 15) shows in detail how the fan-like wing of an earwig is closed up. The housefly makes use of such a line of weakness when it wants to 'feather its propeller' as quickly as possible in order to alight.

A further refinement needs to be mentioned. Most insects have two pairs of wings. Whereas the two pairs of wings of dragonflies and locusts are not connected together and beat alternately, this is not so in Lepidoptera, bugs (Hemiptera) or flies (Diptera). There are many ways of fastening the two wings of one side together with a firm link, and two examples are shown in Fig. 7. One half of the coupling is on the leading edge of the hind wing, and the other is on the trailing edge of the fore wing. The two halves may be hooked together before the insect takes flight, and unhooked again after it has alighted. The adjective 'elegant' would not be inappropriate, having regard to the precision with which the two halves match each other. The linking mechanism may take the form of a row of bristles snapping shut like an automatic coupling on a train. Many wings have a sense organ near the base, or out on the membrane, and this doubtless has some function in regulating flight movements.

Insects that are incapable of flight may nevertheless have wings (Fig. 8), but either the wings are too small or the insect is too fat! Either way the wings are incapable of lifting the body off the ground. This often happens in parasitic insects.

Thus the wing of an insect is an instrument of complex construction. In many kinds of animals the wings can be used

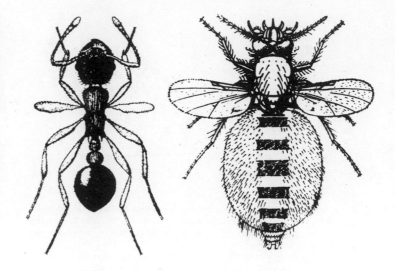

Fig. 8. Examples of insects that have wings but are unable to fly. Left: an aberrant worker of the ant Myrmica scabrinodis *Nyl., with short wing stumps that are useless for flight. Normally worker ants have no trace of wings. Right: the fly* Phasmidohelea wagneri *Séguy (Diptera, Ceratopogonidae, female) is an ecto-parasite of the stick insect in Central America. The wings are one millimetre long, and functional, but they are no longer capable of lifting the heavy, inflated abdomen of the female fly.*

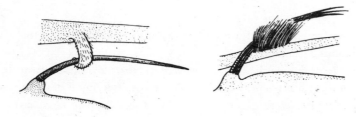

Fig. 7. Two types of coupling mechanism between fore and hind wings in Lepidoptera. Left: hook and single bristle in a male. Right: hair tuft and pencil of bristles in a female.

in two different ways: either to beat actively or to be held stiffly outstretched, when they act as gliding planes. The second case is much the simpler to study, so we will begin with that. Anyone who can remember how a paper dart glides will understand how the gliding principle applies to animals.

Some time ago on a hot day in high summer, I was sitting on the rocky shore of the Adriatic, at Rovigny in Yugoslavia. A full hour of diving and snorkelling had cooled me off considerably, so that I was shivering even in the hot midday sun, and I lay soaking up the warmth. A sea-breeze blew strongly up the rocky slope of the shore, making leaves and pieces of paper swirl over the ridge. Half-a-dozen of the sail butterfly, or scarce swallowtail *(Iphiclides podalirius)* were playing a few metres from where I was lying, and at first I was merely amused by them, but gradually began to watch them more and more closely. In the end I studied them with intense concentration. Light as thistledown, they glided in the wind for several metres with hardly any loss of height. If they headed down the beach into the wind they rose high into the air. Usually at this point they began to flutter, turned in a wide spiral, and returned towards the land. When they passed into the shelter of a tree, a bush or a big rock they resumed gliding flight, and sailed away towards the sea to repeat the performance. They looked like little multi-coloured paper gliders, and behaved like them too; never a stroke of the wings during a glide of ten metres or more, wings held in a slight V-shape, slightly swept back, gliding straight ahead with absolute confidence. They were like small aeroplanes with the engines cut out, in equilibrium between their own weight and the lift of the air stream. Their velocity relative to the wind varied from one to, at most, two metres per

second. Sometimes they reached the ridge of the beach and hovered over one spot for a half, or a whole, second; the butterflies were actually gliding forwards and downwards during this time, but meanwhile the updraught on the ridge was carrying them backwards and upwards, and so for a brief period they were poised at rest. Their approach to a landing was incredibly elegant. By some means they contrived to hover stationary as close to the ground as possible, often without the aid of a single stroke of the wings. Then they stretched out all six legs, and they were down.

The whole process continued smoothly, troublefree and elegantly. Occasionally during their glide they rolled slightly about a longitudinal axis, doubtless because of wind eddies, but normally they were as steady as a rock. I never saw one stall, or get into a spin, in the way a badly trimmed aircraft does if its angle of attack becomes too steep in relation to the angle of glide.

These butterflies spent fully 80 per cent of their flight time gliding passively through the air, and their wing movements never lasted more than a brief period. I watched them in comfort for a long time. In the afternoon of the same day I concluded with some experimental observations on the gliding flight of these same butterflies. The laboratory biologist ought occasionally to desert his instruments and go outdoors to observe—and observe again.

5. Gliding and soaring are flying without muscle power

The sail butterfly is the most accomplished of insect gliders and soarers, but gliding can be easily studied in any of the swallowtails. The cabbage white and the peacock occasionally glide for short spells, but only for a bare metre or so. The big dragonflies, especially massive species of *Aeschna,* are also accomplished gliders: they do not often glide, but when they do so they can keep it up for a surprisingly long distance. It must be quite a feat for them to achieve the right balance between their two pairs of wings, which are not locked together but rather are held well apart in flight. Ordinary field grasshoppers can glide, and the desert locust is a master of sustained gliding, having been observed to cover two-and-a-half metres without a single stroke of the wings.

Someone once saw a swarm of locusts soaring in a thermal for six hours, circling round and round, but never beating their wings. The desert locust can climb to great altitudes by this means, and has been seen at 900 metres above ground level. Does it perhaps return to earth from this height in one continuous glide?

It is interesting to note that typical gliding and soaring flight, with the wings outstretched, do not generally require any effort from the insect. No energy is expended, either because the wings themselves automatically take up the gliding position as soon as the flight muscles are relaxed—as in the big dragonflies—or else there is a 'click mechanism' which puts the wings into the right attitude mechanically.

This latter method is how butterflies assume a gliding attitude.

The fore wings of the desert locust have a particularly positive regulatory mechanism, which is not triggered off until there is an airflow of 64 km/h over the body. Since in ordinary flight the locust seldom exceeds 16 km/h there is a built-in safety factor of four times. A similar safety factor is provided, for example, in lifts, though here it is usually at least ten times. Thus if the lift is labelled 'maximum load 300 kg' it can be loaded up to 3000 kg before the cables will part.

Many midges, stag-beetles and hoverflies also have regulatory mechanisms of this kind. One thing is certain: the ability to fly without muscular effort is by no means a rare accomplishment among insects.

Up to now we have talked mainly of 'gliding' and said little about 'soaring', but there is no essential difference between them. Gliding and soaring both describe flight without any special means of propulsion, whether it may be muscles or motors. The two forces involved are gravity and aerodynamic lift. It is usual to apply the term 'gliding' to non-powered flight in still air, and 'soaring' to non-powered flight in moving air, such as the 'thermals' of rising air that occur over hot ground, or in the updraughts caused by wind blowing against a steep hillside.

When a buzzard, an eagle or a vulture circles in the column of a thermal, rising higher and higher all the time, it is soaring. This could be correctly expressed in different words by saying that the bird is gliding downwards in a rising column of air, gliding being always downwards, since it takes place under the force of gravity. Suppose that the warm air is rising at a rate of three metres per second and that during the same second the buzzard descends one metre in its glide, then the resultant effect will be that the buzzard will soar at a rate of two metres per second (Fig. 9). This is obvious enough, yet it seems to have remained a mystery for a very long time, right up to the present century. This phenomenon of soaring flight can be demonstrated very easily. It is only necessary to blow gently from underneath against a wisp of cotton wool while it is sinking down towards the floor. As soon as the updraught is greater than the falling speed of the cotton wool, the latter will remain poised, and by taking a little trouble it can be held there. A buzzard hovering poised in a thermal is in a similar equilibrium. If the thermal is too weak to support the buzzard, the bird loses patience, and flaps its way to some other thermal that gives it better support.

A glider pilot does the same thing. He climbs steeply in one thermal, then leaves this and glides obliquely downwards towards the nearest thermal, in which he soars again, and so on. By this means he can make long cross-country flights without having any engine. The only snag is that each time he must find another thermal before his downward glide has taken him to ground level. If he does not, he must land.

If everything is so simple it ought to be possible to make a glider out of an old postcard. Throw one up into the air and it will spin, tumble over and over, or sideslip to and fro down to the floor. Suppose we give the card ballast in the form of a paperclip in the middle of each long side. Now either the nose will rise steeply, or the card will dive towards the floor. If the former happens we move the clips a little forwards; if the latter we move them slightly back. By trial and error we should eventually be able to balance the postcard so that it will maintain a smooth steady glide. Now the paperclips have not improved the flying ability of the postcard, except by providing a means of trimming it into equilibrium. This is basically all that the sail butterfly has achieved. Is our 'wonder insect' thus no better than an old postcard and a couple of paperclips? We shall see!

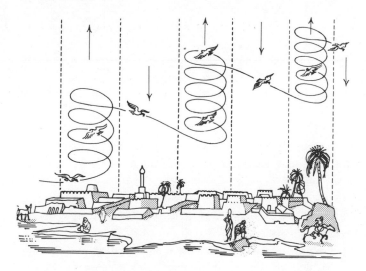

Fig. 9. Over patches of ground that are warmer than their surroundings arise columns of spirally rising air, which are called 'thermals' (indicated by arrows pointing upwards). Birds can ascend in these without having to make a single flap of the wings, and then glide downwards in an adjacent column of air that is stationary or descending (indicated by arrows pointing downwards). In this way birds can cover long distances across country without any active expenditure of energy, as this sketch of a soaring vulture shows. Recent research shows that this apparently simple phenomenon is actually much more complex than it seems; for example, great gusts of air periodically break away from the ground and rise upwards as a bubble.

Up to this point we have spoken of the force of gravity and the aerodynamic forces as being the sole factors determining the glide path, but we do not really know what the 'aerodynamic forces' are. The Greek word for force is *dynamis*, and for air it is *aer*; so aerodynamics is the study of forces due to the air. As you can imagine, the forces acting on a fixed wing are much simpler and easier to understand than the forces acting on a beating wing. So to start with we shall spend a little time considering the aerodynamics of a rigidly held insect wing. In order to understand this better we shall first discuss the problem of forces and their representation.

In physics a force has three properties: magnitude, direction, and point of action. If you raise a bucket of water vertically, you need to apply a force. The magnitude of the force is given by the sum of the weight of the bucket and the weight of the water, the direction is vertically upwards, and the point of action is the middle of the handle, where the bucket is grasped (see Fig. 10). We measure forces in grams and kilograms, but physicists make a clear distinction between kilograms of mass and kilograms of force; we shall keep this distinction in mind without going into the details of the theory behind it, or becoming involved with dimensions.

Weights are also forces, and they may be measured in grams of force. A litre of water weighs approximately one kilogram. Let us assume that our bucket weighs seven kilograms. The force corresponding to this mass acts directly downwards; if we let the bucket go, it falls in this direction. It is the gravitational field of the earth which gives the bucket—and every other mass—its weight, and the point of application of this weight lies at the centre of gravity, in our case approximately at the centre of the mass of water. In this very simple example we have already had to take two forces into account, the weight of the bucket and the force which we have to apply to hold it. The physicist would call it a *pair of equal and opposing forces*.

Another experiment: we attach a large balloon to the bucket. We fill it with just enough gas to make it float, so that it does not rise and does not sink. We call the force which the balloon exerts on the bucket the *lift*. The bucket floats when this force, the lift, is just equal and opposite to the other force, the *weight*. The pair of forces is balanced, as their magnitudes are equal. Their directions are opposite, and they act along a common vertical line. Only the points of action are different; remember that the weight acts at the centre of gravity, and the lift at the centre of the handle. In fact the centre of lift is even higher, at the middle of the balloon. This

does not matter as long as the lines of action of the forces are the same, the point of action of the lift being exactly above the centre of gravity. The bucket is then in stable equilibrium, and cannot be tipped over.

The physicist likes to draw diagrams of these relationships. We should follow his example, as it makes the situation easier to visualise (Fig. 10). Let us choose for the weight, instead of the water, a garden gnome weighing seven kilograms, and hang this gnome on to the bucket. Imagine an angel to provide the lift, by standing on a cloud and lifting up the gnome with a rope. Because most of the weight is provided by the gnome, we can assume the bucket itself to be weightless. We could go further, and replace the whole load by a concentrated mass at the centre of gravity, C. Lift and weight act along the same line, so it does not matter how far away the points of action are from each other. We can put a strong thread of any length between the balloon and the bucket. It makes no difference to the relationship of the forces whether the thread be long or short. For this reason we can ignore the distance between the points of action, and leave it out of the diagram. We let the lift act directly at the centre of gravity. Finally, the gnome representation does not look very scientific. The physicist represents the forces graphically with arrows. The length of the arrow represents the magnitude of the force, the direction of the arrow represents the direction of the force, and the point at which the arrow starts represents the point of action of the force. In this way all three properties of

Fig. 10. Considerations about the development of an upward force which is to compensate a weight. Explanation in the text.

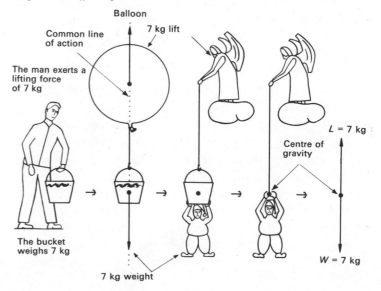

the force are symbolised. Lastly we replace the word lift by the abbreviation L, and weight by W. By this series of steps we have arrived at the usual graphical representation.

This symbolic representation can also be applied to other cases, and it is possible to use it to combine several forces. If instead of the one angel we had two, who stood facing each other on separate clouds, and pulled obliquely upwards, then the diagrams would look like Fig. 11. The two partial lifting forces H_1 and H_2 are combined according to a parallelogram of forces, so that the resultant gives the necessary upwards force. They have to be of such a magnitude and direction that L is again seven kilograms of force, acting vertically upwards.

If this condition is satisfied, then it is only necessary to draw L in the diagram. It is not necessary to know that it is actually the resultant of H_1 and H_2, so they can be left out if L has been successfully constructed from them. It is only

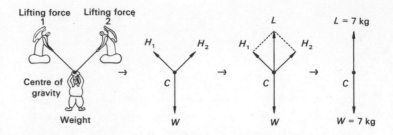

Fig. 11. *A sketch diagram and a physical diagram to illustrate the combination of two component forces to produce an upward force which compensates a weight. Explanation in the text.*

important that a pull of seven kilograms-force is available, and not how this pull is produced. It could be produced in a completely different way. We shall next, as an example, form the necessary upward force by combining so-called lift and drag forces.

7. A little about the aerodynamics of the insect wing

If a butterfly is gliding in still air, it obtains its airflow by virtue of its forward motion. The direction of the airflow is exactly along its glide path, but in the opposite direction, as shown in the sketch of Fig. 12. During gliding flight the butterfly sets its wings at a slight angle relative to the direction of the wind, and therefore also relative to its own glide path. The technologist would say that the wing has a certain angle of attack, which we represent by the Greek letter α. The diagram also shows the angle which the glide path makes with the horizontal. This is called the angle of glide and is represented by β.

Let us suppose for a moment that the butterfly sets out from the top of a high poplar tree, and lands on horizontal ground after gliding in a straight line for a measured distance s. We can call the height of the poplar the 'gliding height' h, and the distance of the point of landing from the base of the tree the 'gliding distance' w. From the height h, the distance w and the glide-path s we can construct a triangle; in the diagram it is cross-hatched. We symbolise the butterfly by its centre of gravity C, and represent the wing by a thick bar. A further point is that the triangle is right-angled, because a poplar normally grows vertically, and its trunk makes a right angle with the ground.

Let us carry out an imaginary experiment. Suppose we were to sit on top of the poplar and allowed various objects to glide away from us: a flat piece of wood, a weighed postcard, a small paper dart, a balsa-wood model glider, and a good full-sized glider. The objects would land at increasing distances from the base of the tree, their glide paths becoming increasingly flatter, and the angles of glide β_1 to β_5 becoming increasingly smaller (Fig. 13).

The flat piece of wood would drop like a stone almost vertically to the ground. On the other hand, the high performance glider might travel horizontally at least 15 times

Fig. 12. *Schematic diagram of the geometry of a gliding object. Explanation in the text.*

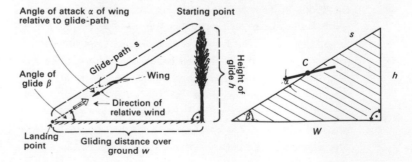

as far as the height of poplar; from a starting height of 12 metres that would be $w = 12 \times 15 = 180$ metres. The other values lie somewhere between these two.

From this example we have obtained a way of measuring the efficiency of a gliding machine in terms of the gliding distance for a given constant gliding height. By this standard the carefully folded paper dart is better than the postcard, because it can glide further. This can be expressed very easily mathematically by the cotangent of the glide angle:

$$\cot \beta = \frac{180}{12} = 15.$$

The cotangent of β is also called the glide number, so that a higher glide number indicates a better gliding performance. Our high performance glider has a glide number of 15, and this is quite a high figure. The postcard might possibly achieve a glide number of 1. For a height of 12 metres it can only glide 12 metres, and in fact it tends to strike the ground at an angle greater than 45°. A glider with such a low glide number could only crash straight into the ground.

The introduction of the angle β and the glide number has many advantages from the computational point of view. It is not necessary any longer to run around with a tape, measuring heights and distances. It is possible to dispense with h and w altogether, since it is only necessary to know the angle β and this can be obtained, for example, from a double exposure photograph taken with an exactly horizontal camera. But even for this experiment it is necessary to have a correctly proportioned model capable of flight, since it is not possible to measure the glide angle of a single wing by itself. Moreover, this method is not suitable for a series of experiments because of the problem of making every launch exactly the same; gusts of wind may distort the picture, and so on. Instead of having a moving model in still air, it is better to keep the model still, and blow a constant-velocity stream of air past it with a suitable apparatus.

This is the principle of the wind tunnel, which is in common use in the aircraft industry. Here the model is held quite still in a measuring chamber, and the experimenter can make all his observations in comfort. Of course, it is no longer possible to measure the glide number directly, since there is no launch position and no landing point, and so this must be calculated indirectly by measuring the aerodynamic forces.

So we have got around to the aerodynamic forces at last. Let us go straight back to the butterfly, and see what forces there are and how they operate. The butterfly has a weight, which acts vertically downwards, and is represented in our next and final sketch (Fig. 14, right) by the arrow W. At every instant during gliding flight, W must be balanced by an equal and opposite upward force, so an arrow U of equal length pointing upwards has been drawn. There is no

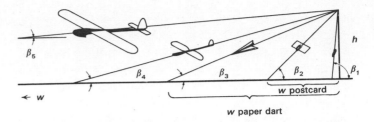

Fig. 13. The glide angle β and the gliding distance w for various objects, which are launched from the same height h, see text. $\beta_1 = 85°$, $\beta_2 = 45°$, $\beta_3 = 23°$, $\beta_4 = 16°$, $\beta_5 = 5°$. At the present-day powered aircraft achieve a glide angle of 5°, and special high performance gliders easily do better than 2°.

difference between this and our previous example of the balloon. It does not matter whether the object under consideration is stationary or gliding at constant speed, so long as the force U is developed. We do not have to have a balloon!

Every driver can do the following experiment. Get someone else to drive you along at about sixty miles an hour and put your flattened hand a little way out of the window, rotating it slowly on its longitudinal axis. When it is horizontal, so that the angle of attack is zero, a slight resistance can be felt, the air stream trying to push the hand backwards. As it is rotated, a considerable upward force is experienced, because the hand has been given a positive angle of attack α. However, if the hand is turned with its forward edge downwards, then the force is also directed downwards, because the hand now has a negative angle of attack. The resistance and the lateral pressure are aerodynamic forces, and are always present when a stream of air impinges on an inclined plane. In every case the resistance is along the same line as the wind, and the lateral force at right angles to it.

So we can now draw into our diagram a drag D in the direction of the glide path, which is the same as the direction of the air stream. We draw the lift L at right angles to this line. The lengths of these two lines must be such that they compose a parallelogram of forces which gives the required upward force U. The principle is the same as with our example of the angles. From the two components D and L we have derived a resultant U which is just that required to balance

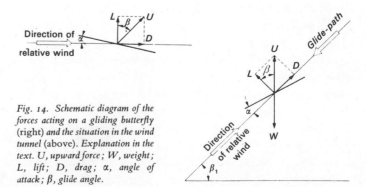

Fig. 14. Schematic diagram of the forces acting on a gliding butterfly (right) and the situation in the wind tunnel (above). Explanation in the text. U, upward force; W, weight; L, lift; D, drag; α, angle of attack; β, glide angle.

the weight. It may seem a little strange that the drag also makes a contribution to the upward force, but this is in fact so. The relative magnitudes of D and L depend on the angle of attack α, and a gliding animal will always adjust its angle of attack so that the resultant force exactly balances its weight. For the sake of completeness it should be mentioned that the forces have been given various different names in the text books. Our upward force U is sometimes called the resultant force R.

Now we know what aerodynamic forces are, and we also know that by measuring them indirectly we can work out the glide angle. Perhaps you remember the geometrical rule learned at school, that if lines are drawn perpendicular to the two lines forming an angle, then the two new lines enclose the same angle as the original one. The glide angle β is formed by s and w. L is perpendicular to s, and U is perpendicular to w. Consequently, the angle included by L and U is also β. The cotangent of β is not only w/h, but also L/D. Our high performance glider has a glide number of 15. Spelled out, this means that gliding distance divided by gliding height = lift divided by drag = 15 divided by 1. So instead of measuring the distance and the height, we measure the lift and the drag. We simply divide the first by the second, and that gives use the glide number.

8. Painstaking experiments mean safe aircraft

Before you begin to be bored with too much theory, let us have a look at some experiments. Our immediate problem is how to measure the components L and D for a gliding butterfly and this is the kind of tricky problem that is always appearing in modern experimental biology. The principle is often simple, but the experimental work may be difficult. In our case we simply take two spring balances, one for L and one for D, hang a dead butterfly with outstretched wings between them, and blow on it with a wind tunnel. The tunnel would have to point along the glide path, that is making an angle of β with the horizontal. This would not be very convenient, and it is better to rotate the whole system through an angle β, so that the wind tunnel is horizontal. This rotates the force components as well, so that L now is acting vertically and D is acting horizontally. The relationship of the forces is not altered, as is shown in the upper part of Fig. 14. Plate 8 contains a picture of such an experimental apparatus.

We are no longer concerned with the weight W because the butterfly is suspended by a system of balances and cannot fall, even if the aerodynamic forces are too small. We switch on the air current at a speed of 1.5 metres per second, measure L and D at various angles of attack α, calculate for each α the glide number L/D, and see which α gives the highest glide number. We find that for the butterfly, the maximum glide number is 4, and it occurs at an angle of attack of $\alpha = +12°$. If it launched itself from the top of our 12-metre high poplar, it would reach the ground at a distance of 4×12 or 48 metres from the tree, which is at a glide angle of $14°$. However, it would only reach this distance if it kept the angle of attack at exactly $12°$. If the angle was more or less than $12°$ the butterfly would travel a shorter distance.

Now we can answer the question which we posed in the introduction, about whether a butterfly is efficiently constructed as a natural glider. We find that the glide number of the butterfly is better than that of the postcard, the paper dart and the balsa-wood model, and is exceeded only by that of a well-built glider. A high performance model glider seems to make the butterfly look poor by comparison, but the comparison is not fair because the butterfly and the model glider are of different sizes. Because of what, in technical terms, are called the aspect ratio and the effect of scale, it is not possible for small flying machines with rounded wings to have as good gliding properties as larger ones with long slender wings. If these factors are taken into account, the performance of the butterfly is nearly the best which is theoretically possible for a machine of its size. Nature has constructed an almost perfect glider.

Let us look a little more closely at the apparatus. The wind tunnel is so constructed that the air stream which leaves its jet is of constant velocity and is completely free of vortices and turbulence owing to precisely adjustable flow direction controls, turbulence nets and smoothing sections. Balances with which the two components of the aerodynamic force can be measured are called, reasonably enough, two-component balances. For measurements on a butterfly's wing they must be extremely sensitive, because the forces are so extremely small: the thousandth part of a newton, a gram of force, must be clearly measurable. My balance works by making the two force components extend two fine helical

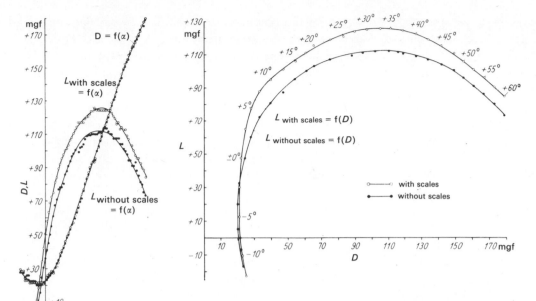

Fig. 15. The resolved forces (left) and polar diagram (right) of Agrotis, a medium-sized noctuid moth. Measurements were made on the whole insect, the wings were spread out in the gliding position. The airspeed was 1.5 m/s. The values of lift (L) and drag (D) are here given in absolute units, milligrams weight (mgf). On the polar diagram the angles of attack are given. The white points indicate a 'normal measurement' (with scales). Before the black points were measured, the scales on the forward half of the upper surface were carefully removed. The drag curve D = f(α) is not altered by this procedure, and its minimum value is also unchanged. The lift curve L = f(α) lies below the one with scales intact, and consequently the polar diagram L = f(D) is also less favourable. The effect of the scales is therefore to increase the lift.

springs by a small amount, a light pointer showing the displacement. The pointer is then wound back to zero by hand and two vernier micrometers used to read the displacement. Then the angle of attack is changed automatically in response to the touch of a button, a heavily geared down miniature electric motor altering it by one degree in every ten seconds. The angle of attack is measured on a circular scale using a telescope. The measurements are repeated, the angle is changed again, and so on, at intervals of one degree, from -15 to $+60°$. One series consists therefore of 75 measurements of lift and 75 measurements of drag. After every experiment the balance is recalibrated, using the smallest possible weights connected over extreme precision roller bearings to measure the vernier displacement as a function of the load.

The components of the aerodynamic force have now been obtained in terms of a displacement, and they can be drawn as a curve on a graph. About a thousand arithmetical steps are necessary to arrive at such a curve, because all sorts of corrections have to be made.

What do these curves look like, and what do they tell us? Fig. 15 shows some examples. The angle of attack α is plotted horizontally, and the measured forces vertically. The aerodynamicist calls these resolved component curves, and we must examine them section by section, starting with the curve $D = f(\alpha)$, that is drag expressed as a function of the angle of attack. The expression 'as a function of . . .' means in this case 'what happens to the drag of the butterfly when the

angle of attack is increased in steps of one degree?' For example, we see that at $\alpha = -15°$ the resistance is 30 mg of force, at $\alpha = 0°$ it is about 20 mg, at $\alpha = +30°$ it is 100 mg, and finally at $\alpha = +60°$ it is 180 mg. So the resistance decreases, passes a minimum at $\alpha = 0°$ and then rises fairly linearly. This is easy to understand, because at $\alpha = 0°$ the butterfly presents the least frontal area to the air flow. As the angle of attack is increased, the area presented to the air stream is increased, whether the angle is positive or negative. At $\alpha = +90°$ the wings are at right angles to the wind, and so they naturally generate the greatest resistance at these points, but the butterfly never turns itself at such an angle. The remarkable thing is the small value of 20 mg of force at $\alpha = 0°$. This is about half the weight of a postage stamp!

Now let us look at the lift. This is negative at negative angles of attack, in other words the force is acting downwards, and obviously the butterfly cannot fly in this attitude. At a little under $\alpha = 0°$ the lift is zero. At greater angles it is positive, and rises approximately linearly until $\alpha = 30°$. After that the curve bends and falls away.

This curve is interesting. It tells us that increasing the angle of attack at first gives us greater lift and hence a greater upward force, but that above about $\alpha = 30°$ this is no longer true. In fact, quite the opposite, the lift falls off sharply if the angle is increased further. Flying in the region to the right of $30°$ becomes precarious because the flying body may drop suddenly as a result of rapidly deteriorating lift, and in the vicinity of the ground this causes a crash. The highest angle of attack which it is possible (but foolish!) to reach by pulling on the control column is called the critical angle of attack, and continued flight after passing this critical angle is called 'supercritical'.

Both the lift L and the resistance D contribute to the resultant aerodynamic force (see Fig. 15). We must therefore compare the two curves point for point if we wish to obtain a true picture of the behaviour of an insect or aircraft wing, and this is difficult. It would be much better if there were just one curve which told us about the aerodynamic properties of the wing. We can do this plotting the lift as a function of the drag, $L = f(D)$. This representation is called the polar diagram, and is frequently used as a convenient summary of the characteristics of a wing or of a complete aircraft.

The right-hand section of Fig. 15 shows the same information as the resolved curves, but in polar form. The highest point of the polar diagram is the highest possible lift, and indicates again the critical angle of attack. Beyond this point the polar diagram falls again; with increasing resistance—that is also with increasing angle of attack—the lift becomes less.

Now let us compare different polar diagrams, and see what information that can give us. Fig. 16 shows the polar diagrams of a subsonic aircraft, a locust wing and the minute fruit fly *Drosophila*. The curve for the locust is very similar to that for a butterfly, which we have already seen. The actual forces on a jet fighter and a fruit fly wing cannot be shown together on the same diagram as they differ by a factor of many millions, so instead of showing the forces themselves we plot their so-called coefficients, that is the coefficient of lift C_L and the coefficient of drag C_D.

I will not bother you with an exact definition of the coefficients. It is sufficient for our present purposes for you to note that by using coefficients we have dispensed with actual lengths, forces and velocities, and so wings of very different size and at very different flight velocities can be compared directly. We shall only discuss three points about Fig. 16, but these suffice to tell us a surprising amount.

First of all the maximum lift $C_{L\,max.}$. It is very important, because the larger it is the larger is the upward force on the gliding wing, and the larger the load it can carry. For the aircraft, the locust and the fruit fly, we find that $C_{L\,max.}$ is 1.5, 1.1 and 0.85 respectively. Insect wings are therefore worse than artificial ones.

Now we wish to see at what angles of attack these values of $C_{L\,max.}$ occur. In the same order, we find $\alpha_{crit.} = 19.5°$, 25° and 50°.

So the critical angle of attack is much higher in insects than in aircraft. We know that on the one hand there is only a noticeable lift when α is greater than zero, and that on the other hand the dangerous region of supercritical flight lies to the right of $\alpha_{crit.}$. Aircraft and animals can therefore fly only when the angle of attack is between about zero and $\alpha_{crit.}$. And we can see that for the artificial wing this region is very small, less than 20° for the aircraft, whereas in nature it is much larger, 50° for the fruit fly.

Our polar diagrams have shown us that the aerodynamic forces are especially large at high angles of attack. These forces are needed particularly for landing, so at this time the angle of attack must be high. In this respect the insect wing has an enormous advantage. Pilots have to be very careful, they can easily stray into the critical region which occurs at fairly low angles of attack, and then drop suddenly and crash. On the other hand, butterflies and locusts can easily obtain the high forces, because they can tip their wing up to 30° before entering the critical region. Fruit flies can manage 60° and more.

Finally we can obtain still more information from the form of the curve beyond $\alpha_{crit.}$. For the aircraft wing it drops off rapidly in this region, and if an aircraft goes beyond $\alpha_{crit.}$ near the ground, that is the end of it. The lift diminishes catastrophically and the aircraft falls like a stone.

The butterfly also enters the critical region if the angle of attack exceeds $\alpha_{crit.} = 30-35°$, but because the polar curve is flatter, the lifting forces decrease more slowly, so it has time to make correcting movements, and the danger that it will go into a spin is less. The little fruit fly is even better equipped because its polar diagram has a flat plateau and falls only slowly even at very big angles. The fly can do more or less what it wants to; it is not able to stall because it has no critical region, and so it is completely stable in flight.

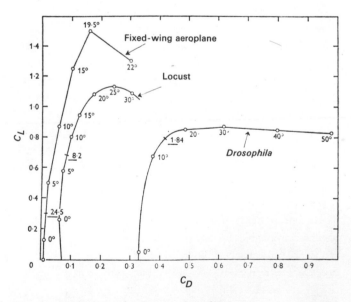

Fig. 16. Aerodynamic polar diagrams $C_L = f(C_D)$: a conventional aerofoil (NACA 2049) with an aspect ratio of 6 at a Reynolds number 5×10^6; the hind wing of the desert locust Schistocerca gregaria at a Reynolds number $Re = 4 \times 10^3$; and the wing of the fruit fly Drosophila virilis at $Re = 2 \times 10^2$. The angle of attack in degrees is given next to the points. On each polar diagram the point of maximum lift/drag is indicated by a short perpendicular line, and the absolute value of this ratio is given underlined next to it. Note the following changes which occur in the transition from the aircraft wing to the wing of the fruit fly. The polar diagrams are shifted to the right (greater drag!), the plateau of the diagram becomes broader, the values of C_L become smaller, but larger relative to their maximum at smaller angles of attack, and the maximum lift/drag ratio becomes less.

We see therefore that insect wings generate high lift and are safe from stalling. Since excessive angles of attack most frequently happen as a result of a sudden gust of wind, we may say that insect wings are largely insensitive to gusts. In this respect they have definitely improved upon manmade aircraft.

We still need to be careful about comparing nature with technology. We have already seen in our discussion of the glide number that, on purely theoretical grounds, longer and slender wings should give better glide numbers. Here the influence of size works in the opposite direction. Theory suggests that small and slowly moving wings should have better polar diagrams, but this effect alone is not sufficient to explain the large differences which we have found. Insect wings are additionally specially constructed to give high lifts at high angles of attack in safety.

All the constructional details which we learned about before contribute to this end: rows of bristles, hair, scales, corrugations, arches and veins. These fine structures affect the layer of air which is in direct contact with the wing and which measures only a few tenths of a millimetre thick, the boundary layer. This extremely thin layer has a decisive importance. All the aerodynamic properties of the wing depend on how the boundary layer moves, and so one can say that every hair and every scale on the wing of an insect has its function. In this field Nature does not make anything superfluous.

Therefore if, for example, the butterflies' scales influence the lift beneficially, the polar diagram would become worse if the scales were to be removed before the measurements were made. The hypothesis can be stated thus: scales serve to increase the lift. Only measurements can show whether this hypothesis is correct or not. In my laboratory a long series of experiments has shown that it is indeed correct. The right-hand part of Fig. 15 shows the polar diagram of a noctuid moth with and partly without scales. First the wing was measured normally. Then it was turned back to $-15°$, the scales were removed from the first third of the upper surface of the wing, and the measurements were repeated. The lift was less at every angle of attack.

Now you have some insight into the deliberations and the logical development of a programme of research. I very much wanted to show you how critically one has to argue in science. If I were simply to say, 'The insect wing is better than the aircraft wing, so let's build a large insect wing and hang an aircraft body on it', that would be false and misleading. The insect wing is indeed 'better' than the aircraft wing, but only in certain respects. This advantage is only obtainable over a certain range of sizes, determined by the static properties of the wing, and then only because of the special constructional materials that are available in the insect world.

One must therefore be very careful in comparing biology and technology.

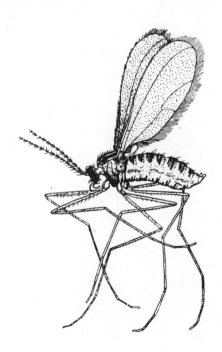

9. In the realm of thousandths of a second

The observer of nature is not only thrilled by the elegant, silent glide of the sail butterfly—he is also excited by the buzzing and humming of the darting blue-black carpenter bee. This tiny, droning machine, what a miniature bundle of energy it is! If you pick up the bee between thumb and finger you can scarcely hold it; the whole body of the insect vibrates so hard as it struggles to escape. Just as impressive is the flight of those fighter planes the hornets and horseflies, those helicopters the bumble bees and the spurge hawk moth. Their speed is by no means insignificant, the distance they cover is amazing—and their lifting power is enormous.

The wasp *Philanthus* is only two centimetres long, but it can carry a honeybee through the air, holding it below its own belly as shown in the photograph in Plate 21. The bee does not weigh much less than the wasp that is carrying it. Solitary wasps, slender graceful insects, can carry bulky caterpillars or spiders hundreds of metres through the air (Plate 19).

This ability of insect wings to cope with such burdens is an impressive feat of strength. How can we begin to understand it? 'Kineo' in Greek means 'I move', and the kinematographic film gives us the means of studying pure movement. In the present context we are concerned with the kinematography of the wing beats of insects. For a start let us confine ourselves to making visible the downbeat only, and trying to understand what goes on. We are not even concerned as yet with how the wing exerts pressure on the air, not how the muscles operate the wing on its hinge with the thorax. We are simply trying to observe and explain what the wing actually does when it is set in motion, how it moves, twists, flexes and warps. Once we have described the movement we can then go on to ask the question: 'Why does the wing move in this particular way?', and also 'What aerodynamic function does this movement serve?'

Let us take as an example that commonest and most mundane of insects, a fly. A blowfly, perhaps, one of those yellow-green or deep blue metallic buzzers. Flies have one great advantage for this kind of investigation—they have only one pair of wings! This simplifies both experiment and subsequent description.

So let us pick up a fly, holding it firmly in our fingertips, and see exactly how it buzzes. We can see two glistening semicircles, and nothing more of the wings. If we look at the fly in a side light against a dark background we may see flashes of light, or crescent-shaped clear patches in the shining arc. This overall impression is as much as we shall get because the wing beat is too rapid for our eyes to follow. We can try to persuade the fly to beat its wings more slowly, perhaps by mischievously loading the wing with a tiny scrap of paper, but all to no avail. The fly refuses to respond. Either it beats its wings as before, or it stops altogether, so we get nowhere.

Since the fly refuses to beat slowly for us, we in our turn must 'look more quickly', as we can do with the aid of that convenient slow-motion invention the high-speed cine camera. This miniature cine camera uses 16-mm film, and can take up to 8000 pictures in a single second. A 30-metre spool of film runs continuously not intermittently, and at the speed of an automobile. If the processed film is projected at the usual speed of 16 frames per second, the action is slowed down 500 times (8000/16 = 500). Special analytical projectors can operate as slowly as two frames per second without flicker, and so achieve a slowing down of 4000 times (8000/2). At this speed an action that really occupies only one second takes 4000 seconds to project, and that is over an hour! The fly beats its wings about 200 times a second, so that one up-and-down stroke takes one two-hundredth of a second in life, and 20 seconds (4000/200) when projected. A further convenience is that the projector can be stopped and a single exposure studied for any desired length of time. In the course of my research I have made drawings of hundreds of such single exposures, and analysed them graphically afterwards. I had already been working for two years before I had the first analytical film in my possession, but that amount of time was necessary to work out a reliable analytical technique. We shall soon see how essential this technique was.

Plate 3. Above: the small earwig, Labia minor, has extended its right wing and is about to unfold its left wing with the aid of its abdominal 'pincers'. It will be ready to take off in about 1 to 2 seconds. Below: the longhorn beetle Rhagium mordax is ready to take flight. The wing cases are already raised high, and the final fold at the tip of the membranous hind wing is just straightening out. Take-off may follow in ½ to 1 second from now.

Plate 4. (Over page, left.) Two separate stages in the downstroke of the wings of a North American cicada. Above: the start. The leading edges of the wings have already started to move downwards, while the trailing edges are still completing the previous up-stroke. The wings are thus strongly twisted. Below: in the middle of the downstroke.

Plate 5. (Over page, right.) Continuation of Plate 4. Two separate phases of the upstroke. Above: in mid-stroke. Below: towards its end. The forewings are curved towards the middle axis of the body, and the small hind wings beat in phase with them. For these pictures the cicada was attached by the abdomen to a fixed needle. Pictures taken with a special electronic flash unit which gives a flash duration of one millionth of a second.

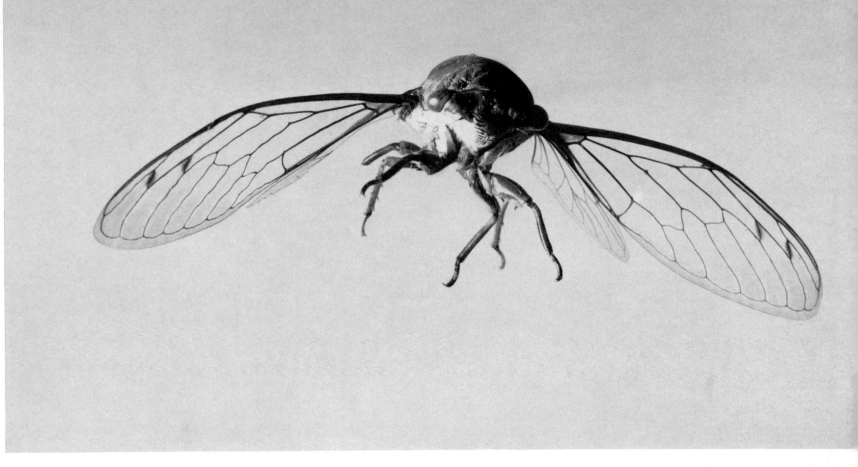

It is not so easy to film a fly. You could hold the camera in your hand and run after the animal when it flies past, but who would try to do that! Some clever people have hit on the splendid solution of fixing the camera and holding the insect in front of it, attached by the thorax or abdomen to a piece of wood which is held in a stand. As soon as its legs are removed from the ground, the fly begins to beat its wings, and the shutter can be pressed. Very good. We have a beautiful film, but does it tell us anything about how the insect would have moved its wings in free flight? The experimental conditions are unnatural, but does this make much difference?

A moment's thought about the aerodynamic forces will give us the answer. We are stipulating that the insect should fly horizontally without accelerating, that is, at constant velocity. We want to call this the normal flight attitude. What forces are acting on the animal in free flight? Firstly there is the weight W, and this always acts downwards. The insect must balance it by an exactly equal and opposite upward force L. Secondly, the air resistance D acts backwards. The insect must balance it by generating an equal forward thrust T. So we have two pairs of forces, W–L and D–T, and each must be equal to its partner (Fig. 17). This is valid for normal flight. However, if the fly is held in a fixed position then these relationships are fundamentally altered. The animal no longer has to compensate for its weight because the stand is carrying it. There is no resistance either, since it is not moving through the air. The fly can generate any lift and thrust that it wants to. It can move the wings in any way it wishes, or leave them quite still—nothing can happen, the animal will remain fixed. If we film it, we don't learn anything, because we do not know what the animal was trying to do. Perhaps it was moving its wings in such a way that a large lift was generated, so that it would have ascended sharply if it had been released. Maybe the opposite was true. We have no idea, because we cannot measure these

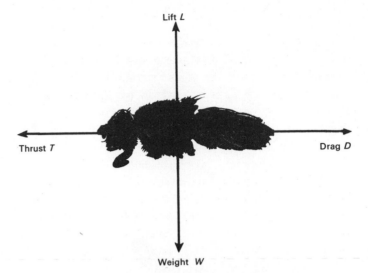

Fig. 17. *The forces acting on a flying insect. Lift and weight, and drag and thrust, must balance during steady horizontal flight. It is assumed that all the forces act at the centre of gravity. In fact this is not the case; because the points of action are not exactly coincident, a moment is generated which could make the flight unstable, and which must therefore be continuously corrected by active changes in the form of the wing beat and passive adjustment of control surfaces on the abdomen and legs. We are not at the moment interested in stability and so the diagram is a close approximation to the actual situation.*

tendencies. The basic idea of fixing and suspending the insect is good, but two more criteria are necessary. Firstly, the insect must be allowed to balance the forces itself as it does in free flight, and secondly the apparatus must be able to detect when this balance exists, so that it can signal 'operate the shutter immediately, at this instant things are just right'.

Let us assume that these conditions are satisfied. What have we achieved? Again two things. Firstly, we can say with certainty that at the moment of exposure the insect was moving its wings in the same way that it would have done if it were flying horizontally at a speed of so many metres per second, even though it was actually fixed. If all the conditions really do remain constant, and during the filming the flight attitude does not alter, then we can also say that each wing beat will be the same as the next. If this were not the case, then the different wing beats would have produced different forces and the measuring devices would have detected it.

The developed films have shown that this condition has been fulfilled exactly, each of these complicated wing beats looking exactly like the rest, just like a machine. This is invaluable from the practical point of view, since the researcher only needs to measure a small number of beat periods. The rest of the film represents an unchanging flight

Plate 6. A ladybird of the genus Coccinella *makes a steep take-off from a willow catkin. The legs are held in a typical attitude. The wing cases beat in rhythm with the wing movements.*

condition. Only when one has fully understood this, the most important form of flight, is it worth while studying the kinematics of unsteady flight, for example take-off and landing, upwards accelerating flight or turning.

Now I would like to give you a description of the techniques that we have used to achieve this in the laboratory, and to tell you a little about the experimental apparatus, so that you can appreciate how long it took to build, and how, by careful planning, a few wing beats—to be more precise, just one wing beat—can be made to tell the whole story. Then we shall want to look in more detail at a single up-and-down movement, to examine one two-hundredth of a second in the life of a fly. It is unbelievable how much can happen in a two-hundredth of a second. The planning and building of the apparatus to answer this question is just an example of everyday scientific work. No scientist can escape this detailed work, whether his problem is as intricate as ours, easier, or much more complicated. Such practicalities occupy most of his time.

We have already seen that a wind tunnel is very useful for the study of gliding flight. For the present problem it is indispensable. Let us imagine that the fly is moving directly into the air stream which is emerging from its nozzle. Let the flying speed of the insect be exactly 2 metres per second, and the wind tunnel also blow with exactly this speed. Seen from our point of view, the insect remains stationary, because, in a unit of time, it flies as far forwards as it is blown backwards by the wind. Theoretically we could now film it. In practice, however, these moments of exact velocity equivalence are too short, and also the fly moves away to the side easily. For this reason the insect is glued at the tip of its abdomen to a fine wire with a drop of insect wax, and the wire is attached to an instrument which might be called a 'force-compensating suspension'.

This instrument fulfils the conditions which we require: it allows the insect to adjust itself to free flight, and signals when this has happened so that photographs can be taken. It also prevents the insect from leaving the field of view of the camera if the flight attitude suddenly changes. A diagram of such an apparatus appears in Fig. 18. In principle it consists of three coupled balances, one for the turning moment, one for lift and one for thrust. It seems hardly possible, but after careful balancing the tiny fly can move the comparatively enormous three-stage balance with the greatest of ease. Obviously it has to be built with great care and be as light as possible.

If you look more closely at the diagram, you will see that the turning moment balance, which is adjusted practically to neutral equilibrium, allows the animal to set its abdomen at any angle to the airflow. For example, the fly can slant itself upwards or downwards since the balance moves with it.

At the beginning and ending of flight such movements are frequent. After this initial phase a period of normal flight usually follows. During it, the insect aligns itself exactly with the direction of the wind.

The turning moment balance is situated right at one end of the long arm of the lift balance. Before the flight the insect is weighed. Suppose it weighs 32 mg. A wire weight of 32 mg is placed on the other end of the balance, and the rider is moved until the balance indicates exactly zero. Then the wire weights are removed, and the balance tips up until it hits the end stop. When the insect is flying and generating lift, it raises the balance again. If it generates exactly 32 mg, the balance will again indicate zero. Now we have the condition of normal flight, at least as far as the pair of forces $W–L$ is concerned; the insect has balanced its weight W by an equal lift L. After the starting up time, the balance sometimes stays at zero for a quarter of an hour, as though frozen. The fly seems to be comfortable when L is equal to W. This is hardly surprising as it is its normal flight attitude. If the pair of forces $T–D$ is also balanced, then the shutter can be operated. The film shoots through the high-speed camera with a deafening siren-like howl.

But we are not finished yet. We must first understand

Fig. 18. Diagram of the three-component (turning moment, lift and drag) force compensating suspension used to support the flying insect in front of the wind tunnel. Each component of the balance is set on knife edges, has its own oil damping, and can be zeroed with adjustable threaded weights. The three axes are drawn with broken lines. The arm which projects into the air stream is shielded with a profiled cover. The contacts are connected to a servo mechanism which adjusts the air speed to the value set by the insect. The inset diagram shows one of the earliest attempts to register the motion of the wings (second half of the nineteenth century). The insect is fixed to a stand, and writes the path of its wings on a piece of smoked paper which is moved past it; this is known as a kymogram.

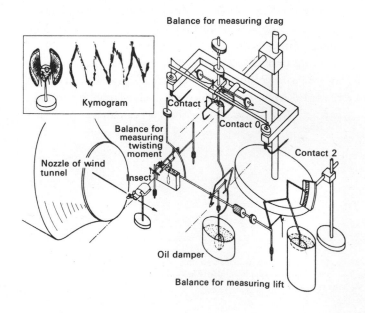

how the thrust balance works. It is in principle a sort of vertical pendulum, but instead of a weight it carries the bearing on which the lift balance can freely move. Let us assume that the pendulum is hanging exactly vertical before the experiment. Now we switch on the air current. It pushes the fly, and with it the whole system, backwards and the pendulum hangs obliquely. Now we 'start' the fly so that it begins to beat its wings and to generate thrust. It pulls the whole system forwards, and the angle of the pendulum is gradually diminished. As soon as the pendulum is exactly vertical again, the thrust must be exactly equal to the drag or backwards force. The force pair $T–D$ is balanced, because the fly—just as in free flight—is generating exactly the same forwards thrust that the wind resistance is absorbing.

The fly chooses its own flying speed. It activates a servo-mechanism which adjusts the speed of the wind tunnel to the value that suits it. It is a very simple feedback mechanism. Assume that the insect wishes to fly slowly. It will reduce its

thrust, but the wind is blowing with the same strength, so it will push the whole system backwards, and the pendulum will hang obliquely. This closes contact 1 (Fig. 18). A small adjusting motor turns a potentiometer, and this reduces the voltage on the wind-tunnel motor, the fan turns more slowly, the wind slows down, and the arm of the pendulum moves slowly back to the vertical. As a result of this the contact is opened again, and the adjusting motor stops. We have now arrived at a new balance $T = D$. The forces are still equal, but they are both smaller than before. Success: the fly is going more slowly, but the forces are balanced as in free flight. If it should occur to the fly to fly faster, the apparatus works in the opposite direction by closing contact 2. The fly can be attached to the previously-mentioned piece of wood or to the balance system. If you watch it flying there seems to be no difference, but the one is a game and the other science. Some people might mock this 'science'. Is anyone really paid a salary so that they can play with flies?

11. 'What a lot of money to spend on a tiny fly!'

All this money, just for a tiny fly! Could not all this time, money and ingenuity have been given to some more worth-while piece of research? Would it not have been more valuable to have studied the aerodynamics of a supersonic airliner—the economics of its operation, a balance-sheet of profits and losses, its commercial possibilities, its value as a money-earner, for cheaper tourist travel, or for faster military transport—than to research into the flight of a small insect? A whole new branch of activity is opening up in the air . . . all very true—but it misses the point of scientific research.

In the first place scientific research is conducted for its own sake and not for any immediate objective. The scientist is searching for knowledge, not for its practical utility but in pursuit of a grand ideal. He is an idealist in the best sense of the word. His research is planned and directed towards a simple goal, which follows on logically from his previous work. He will do whatever may be necessary to fill the gaps that exist in his particular field of knowledge. These endeavours of the research scientist are quite disinterested, and expressions such as 'important', 'significant', 'immaterial' have no meaning for him in their common application. He asks only 'how' and 'for what reason' things happen, and is concerned only whether statements made are true or false.

His objective is to add to knowledge, and in Natural

Science this means discovering general principles which apply throughout the natural world. This is the ideal towards which the dedicated scientist labours, wherever he works, and whatever may be his immediate problem.

It is self-evident, of course, that a scientist cannot always confine himself to this ideal course. A scientist's calling brings with it many other, pressing demands. Nevertheless as soon as he is embarked upon an investigation that may lead to the solution of his particular problem he must, and will, forget everything else, and concentrate on his work, oblivious of the passing of time.

Of course the practical results of applied science are important, too. Nowadays every big industrial plant has its research laboratory. The objective of applied science is not knowledge for its own sake, but, in the last resort, an improved product, or an improved turnover. In universities and technical colleges research is often directed towards practical ends, such as designing a new construction tool or a new manu-facturing process. Seen from this angle there are big differences between one line of research and another. Of course no one can say for certain in advance that some completely academic investigation will not eventually lead to practical benefit. The best example of this is the study of the Foraminifera. These are microscopic animals which often build up a calcareous covering like a snail shell. For a long time the description and cataloguing of hundreds of different kinds of these remained a pursuit of no interest to anyone except students of Foraminifera, but the picture changed with dramatic suddenness as soon as it was discovered that when drilling for oil the presence of certain particular species in the cores may indicate the presence of oil-bearing strata. Today the study of the Foraminifera is of the highest economic importance.

Scientific work cannot be divided into research projects that are of practical importance and those that are not. There is only one, comprehensive, reservoir of knowledge, and every new discovery is something added to this. Every now and then a practical benefit emerges, and the likelihood of this happening is increased the more knowledge is pooled. The biggest industrial concerns all over the world nowadays allow themselves the luxury of a department of pure scientific research. Some large commercial companies employ many hundreds of research scientists and technicians, who follow any conceivable line of research, regardless of whether or not it is directly relevant to the Company's own products. A single lucky discovery by just one of the many people employed may offset the vast expense of years of research. One example of this kind is the transistor, which not only ushered in a new era of electronics, but which also brought a fortune in patent royalties.

12. Two years of preparation—two seconds of photography

Back to our fly. We have 'set it in motion' and it has begun to fly. You will understand the reason for using the expression 'set it in motion' when I describe what happens. Before the experiment the suspended fly is given a piece of paper to hold, which it turns with its feet, but does not let go. After the wind tunnel has been switched on, the paper is jerked away to one side with a pair of fine forceps. The fly has, as it were, lost contact with the ground, and special sense organs at the ends of its front legs signal 'legs free'. This, however, is the situation in free flight, and so the signal means not only 'legs free', but also 'beat your wings'. The fly starts buzzing. This tarsal reflex, as it is called, is as reliable and sure as the starter button of a petrol engine.

So the fly starts buzzing. It swings around a bit at first, but soon aligns itself with the wind, and moves its legs into the correct flight position—front legs pointing forwards, middle and hind legs backwards, as shown in the top drawing of Fig. 46 on p. 116. Slowly the forces balance each other. During this time a red light glows on the switchboard, but as soon as equilibrium is reached, the light goes out, and at last the slow-motion camera can be started.

A completely automatic sequence of switch operations and relay-controlled delays now begins, which switches the right voltage onto the right motor of the camera at the right moment. For one second a current of 150 amps flows through the small 70-volt motor from 20 accumulators: this is over 10 kW, enough power to light a hundred 100-watt bulbs. The film shoots through with an ear-piercing howl. When the last foot is through, the camera switches itself off automatically and the painfully loud screeching dies away.

The fly is not disturbed at all by the noise, and keeps buzzing away quite unperturbed. If it is in good condition, and not too old, it can fly for several hours without stopping, from time to time taking up a new flight attitude. When this happens, the balances adjust themselves, and then remain at the new position of equilibrium. It is important to have a strong light shining down the wind tunnel, to give the fly the sensation of flying towards the light, and the temperature must be high, preferably 27–29 °C, maintained by a thermostat in the wind tunnel. The humidity should not be too low and there are one or two other factors which have to be kept under control.

Once I began an experiment early in the morning with a large *Eristalis* hoverfly, which flew until midday, and was then fed a drop of nourishing 'soup' on the end of a glass rod—a solution of glucose. After that it flew until afternoon coffee time, when it was fed another drop, this time with added vitamins, and then it flew until supper time, when it was fed again—and after that I went home. When I arrived back the next day it was flying along quite happily, and looked as if it had been frozen in its flight position. Had it flown all night? I think it must have done so for several hours at least, since the light had been left on. For every such superfly, there are ten which fly badly for a couple of stretches of a quarter of an hour, and twenty more that don't want to fly at all. Evidently there are lazy elements among the flies as well, so there is no success without patience on the part of the experimenter.

By compensating for the aerodynamic forces, we have achieved a situation closely approximating to natural flight. But another problem is just as important: how should I photograph the fly in such a way as to get the most meaningful results from the films? Should I take it from the side, from the top, or from the rear—or would it be better to take it from all three together. What are we looking for in such a film? Imagine first of all that you yourself are sitting on top of an enormous insect, which is flapping its wings slowly and flying through the air, and that you have to describe the motion of the wings. As you concentrate hard on the pattern of the wing beats, does it matter to you whether the fly is actually moving through the air, or only appears to do so? If the insect happened to be flying in a cloud so that you could not see the background moving past, consciously or unconsciously you would have to take some point on the fly itself as a reference point, not some point in space. But in fact, it makes a big difference what reference point you take. In the first case you see the wing tips describing something approaching a semicircle, or a segment of a circle, with one point of the wings remaining apparently stationary, the basal hinge about which the whole wing turns. This makes a convenient reference point. In the second case (i.e. if you remain stationary and let the fly pass you), you see the wing tips describing a curve with hills and valleys, an extended wave motion. The base of the wing, on the other hand, describes a straight line, provided that the trunk does not swing noticeably up and down. There is no longer any fixed reference point and an arbitrary reference must be found somewhere in space, for example the top of a tree. The insect flies towards this point, passes it, and continues on its way. The observer measures the position of the whole insect and of the wing tips, relative to this point. The appearance of things changes according to the way we look at them, or, in scientific language, the system of relationships varies according to the choice of origin. In the first instance, we had an 'insect co-ordinate' system, because the origin (or point of reference) lay on the insect itself. The second case can best be described as a 'fixed co-ordinate' system, because the origin is fixed somewhere in space.

If we want to describe the motions of the wings themselves, and that is indeed what we intend to do, we should naturally choose the insect co-ordinate system. However, as soon as we wished to tackle the problem of the action of the wings on the air, how they generate lift and thrust, we should choose the fixed co-ordinate system.

If the wings are moving up and down (beat movement), and the whole insect is moving forward at the same time (translational movement), the wings are in a different place at every instant, and they seize a different part of the air a little bit further on than the previous one. Put in a nutshell, if we want to describe just the kinematics we only need to know the motion of the wings; for aerodynamic studies, we must additionally take the translational movement into account. The tip and the base are two especially well-defined points on the wing, so let us study the motions of these two points in an insect co-ordinate system. On this system, the wing base is a fixed point of reference, leaving us with only one problem to solve, the motion of the wing tip around its base.

Unfortunately, the wing tip does not simply swing back and forth like a metronome. Because it is turning as well, it describes part of a circular track, and this track lies at an angle to the body of the insect. Therefore the wing tip moves in three dimensions, backwards and forwards, up and down, and left and right, all relative to the insect, and unfortunately one cannot record all these movements on a single film. In fact, two films would be enough, as we can see by comparing it with photographing a house. Suppose we want to show length, breadth and height. A photograph from a helicopter

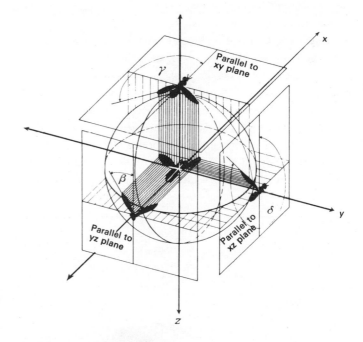

Fig. 19. The projection of a flying insect onto three mutually perpendicular planes (yz, xz, xy), which are perpendicular to the three axes (x, y, z) of an insect co-ordinate system. The path of the wing tip can be described uniquely by the three angles β, γ, δ projected onto these planes.

directly above would tell us the length and breadth, but not the height; a second photograph from ground level at the back would show breadth and height, but not the length. So we now have all three dimensions, but it would be a wise precaution to take a third photograph from the side to show height and length, as a check on the others.

That is just what we do with our insect. We film it simultaneously from three directions, from above, from the rear and from the side, all at a rate of eight thousand frames per second. This is done, of course, in front of the wind tunnel, when the forces are equalised on our balancing system. You know that three co-ordinates are needed to define the position of an object in space, in this case the moving wing tips. It would not be appropriate in this case to use distances, as in the case of the house (length, breadth, and height), but it is much better to work with angles. Look at Fig. 19. Our experimental animal is imagined to be flying in a hollow sphere, against which three mutually perpendicular imaginary planes are placed, and the fly's shadow is projected onto these planes by three rays of parallel light. This gives the same images as those obtained by photographing it from above, from behind and from the side, provided that the camera is set at the appropriate distance. In each picture the base of the wing is a fixed point, from which we can draw two lines, one in a constant direction as a fixed base line, and the other to the momentary position of the wing tip. The angle between these lines changes continually during the wing beat. If we do the same for the other two projections taken at the same moment, we can define the position of the wing tip in space by the intersection of our three lines.

It is obvious that if it is photographed exactly from above, from the rear, and from the side, then the three projections coincide with the three axes of the insect. These are the longitudinal axis, the transverse axis, and the vertical axis. The drawing shows the correct orientation.

In the next picture, Fig. 20, you can see how we have solved the technical problems in the laboratory. For the sake of clarity the whole of the balancing system has been left out of the drawing. The insect is illuminated from the front, from

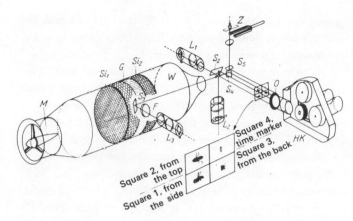

Fig. 20. A diagram of the principle which is used to film an insect flying in front of a wind tunnel along the three body axes. To do this it is necessary to use three low-voltage lamps (L) and five small mirrors (S). A vibrating tuning fork (Z) placed over the objective (O) can also be filmed by the high speed camera (HK). Further abbreviations: G, honeycomb settling section; O, objective; M, motor; Si, sieve; W, wind tunnel.

beneath, and from the right, with sharply-focussed low-voltage spot lamps. For the first of these it is necessary to put a mirror and a window in the wind tunnel. Two of the projections are turned by the small mirrors S_2 and S_4, so that all three now follow parallel paths down the telephoto lens of the slow-motion camera. Every frame is therefore divided into four quadrants. Three of them show the three projections, while the fourth is a time marker which uses the small mirror S_5. This gives exactly one hundred flashes of light per second, produced by a vibrating tuning fork. From this flashing marker one can later determine the exact speed of the film.

Plate 8 (bottom left) shows you an enlarged frame of an original film. The picture from the side appears to be larger, because an additional lens ground down to a rectangle has to be used to get the focus right. This does not matter, however, because we are measuring angles and not distances.

You will understand now why this equipment took two years to build, though of course I was not able to give my whole time to it. Research, like other activities, is unfortunately subject to frequent interruptions.

The experiment is over, and now comes the laborious job of evaluating the results. Miss Christa patiently measured thousands of values of α, β and so on, and a specimen of the results of such measurements is shown in Fig. 21, which represents three successive wing beats. Here the three angles are plotted as a function of time, each point representing the measurement of one angle. You can see that each beat is like the others down to the last detail. The curves have different slopes and bumps, so the wing tip does not move with a uniform periodic motion. It is easy to see from the curves that the forward and downward stroke lasts noticeably longer than the upward and backward stroke. Close examination of the curves reveals a wealth of kinematic detail which does not interest us at the moment. One of these results shows that the stroke plane, the glittering semicircle which can be seen with the naked eye, lies on average at an angle of less than 45° to the long axis of the body. The downstroke leads obliquely forwards, and the upstroke backwards. During a complete beat the wing tip describes a closed curve, which we call the wing tip curve. We have its three components, but not yet the curve itself. There are mathematical and graphical methods for constructing the curve from its three components, and we shall use a simple graphical method.

What we want to do now is to determine the path of the wing tip in the insect co-ordinate system. The base of the wing was the fixed reference point. The tip moves around the base. The distance between the base and the tip, that is the length of the wing, remains constant since the wings don't become longer or shorter when they beat. Any reader interested in geometry will see immediately that the wing tip must move on the surface of a sphere whose centre lies at the base of the wing. (Hang a walking stick on a hook by its handle, and move the other end in every possible direction. The point will describe a section of a sphere. The radius of this imaginary sphere is the length of the walking stick, and its centre is at the hook.) We therefore need a sphere on which we can draw the wing tip curve, and so we buy one of those teaching globes used in schools, which has a blackboard surface and is marked with the lines of latitude and longitude. Every point on the earth is referred to by its longitude and latitude co-ordinates. For example, Munich is at 11.6° east 48.2° north. The co-ordinates are given in degrees, the same units that we used to describe the path of the wing tip.

Imagine a tiny globe with the same radius as the length of the wing in which for example the left wing is beating, and on which it is drawing with the wing tip the form of the wing movement (Fig. 22). You will see immediately that we have stolen our elegant co-ordinate system from geography.

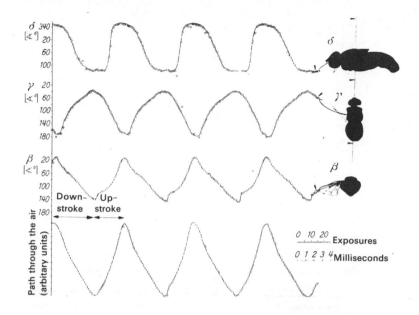

Fig. 21. The three angles β, γ and δ plotted as a function of time during three-and-a-half successive wing beats. (See Fig. 19.) The distance between the points is 1/6400 second. The fly Phormia was flying steadily in front of the wind tunnel, with an air speed of 2.75 m/s. You can see how regular the wing beats are. Points which do not lie on the curve result from a mistake in measuring the film or in the calculations. The lowest curve represents the path of the wing through space, calculated from a path such as that shown on the globe in Fig. 22. It runs in phase with and has a similar form to the β-function.

After applying a correction, our angle β is equivalent to the latitude, and γ to the longitude. Just as the position of a town on the earth is described completely by the latitude and longitude, so is the position of a point of the wing path on the tiny imaginary globe determined by the corresponding angles β and γ. So we read from the co-ordinate curves the values of β and γ for each instant of a complete wing beat, imagine they are latitude and longitude, correct them for the projection, and draw them on our blank globe. In this way we build up point by point the curve we require.

Three-and-a-half successive wing beats give us the same number of three dimensional curves, and these are drawn next to one another in Fig. 22. In order to avoid distortions, the globes are drawn as if they were seen through a telescope from a great distance. The head of the insect is to the left, and the abdomen to the right of each globe (see also Fig. 26). You can see again that the curves follow the same path for each successive stroke. There is also another feature, already present in the component curves, and that is that the downstroke occurs further forwards and the upstroke more to the rear. At no point does the wing tip travel the same path forwards and backwards. The paths lie furthest apart at the centre, and they both make an angle of about 45° to the long axis of the body, as has already been mentioned.

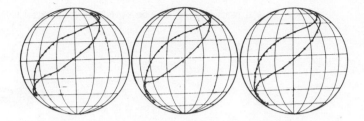

Fig. 22.   *The path of the wing tip during three successive wing beats in the insect co-ordinate system. The paths appear like this if you imagine that the wing tip is moving on the surface of a sphere the radius of which is the same as the length of the wing. The path of the upstroke lies to the rear of the path of the downstroke. The orientation is the same as in Fig. 26; see also the legend of Fig. 23.*

It can be seen already how necessary was the complicated system with the wind tunnel and the three-component balance. If films are analysed which were taken without this technical assistance, the path of the wing tip looks quite different. It is divided at the top and the bottom, but runs into one in the middle, giving the appearance of a long thin figure-of-eight. If this curve is taken as the basis for calculating the aerodynamic forces, strange values are obtained which could never be made to correspond with credible force components. If we had not taken all that trouble over the wind tunnel and the compensating balance system, the results could have been put straight into the waste-paper basket. The whole success lies in the planning of the experiments. In scientific research it is not good enough to do everything in the simplest possible way; all the factors must be weighed against each other and drawn into the planning.

Now let us look at Fig. 23. It shows one of the wing-tip curves in greater detail with all the measured values drawn in. Miss Christa has solved for us the difficult problem of projecting a three-dimensional curve onto a piece of paper— the old and always-troublesome problem of the cartographer, how to represent the globe of the earth on a flat sheet. Here again you have to imagine that the head of the animal is to the left of the drawing, and the abdomen to the right, as in Fig. 26. You see also that every point has drawn through it a line with a triangle at one end. Each of these lines makes a different angle with the path of the wing. The down and forward stroke starts at measurement point No. 1, and runs to point No. 30. Then the upward and backward stroke begins, and this continues until measurement No. 50, which is followed by No. 1 again. Now a new downstroke begins, and so on . . . the whole thing is repeated two hundred times each second. The superimposed lines have a special significance which we must try to understand before going on to the promised chapter about the 'saga of the one two-hundredth of a second'.

Up until now we have behaved as if only the wing tip were important. We have followed the path of just this one point, but there are thousands of other points on the wing. Because the base of the wing is fixed, we do know something else: the path of the long axis of the wing, the series of points between the base and the tip.

Let us take a beer mat and draw a line across it, calling one end of the line the base and the other the tip. The line represents the long axis of the wing, and the mat its spread. Now we hold it firmly between the thumb and forefinger at the base, lay our lower arm on the table, and move the mat around in various ways. We soon notice from our wrist movements that the beer mat moves in two basically different ways. The first is an up and down or backwards and forwards motion. The wrist moves in the same way, but the lower arm remains still. A second way it can be moved is to give it a rotational motion. In this case we turn the hand as if we wanted to drive home a screw. The lower arm rolls around on the table, while the wrist remains relatively stiff. The effect of this is as follows. In the first case the axis of the wing with the rest of its surface swings about but it does not rotate. In the second case the axis remains almost stationary, but the whole wing rotates right and left around it. So we speak in the first case of a beating movement, and in the second of a rotational movement of the wing. You could also say that in the first case it was undergoing a bending oscillation, and in the second case a rotational oscillation.

Now that we have learnt about the two basic types of movement, let us try to combine them. First we move the beer mat obliquely up and down at an angle of 45°, like the wing. Then we add to this movement, which we continue, the turning motion *à la* screwdriver. Whilst the first movement reminds us of beating an egg with an egg-beater, the motion is altered by the addition of the rotation to the sort of movement we would make if we wanted to paint a large figure-of-eight on the wall by just moving our wrist.

You see that movements and oscillations can be combined in such a way that something new is produced. The insect wing also combines beating and rotational oscillations to produce a sort of winging motion. We have already drawn the beating movement in the shape of the wing tip curves and we can superimpose the rotation point for point on the path of the beat. If the wing is viewed from the wing tip along its axis, it appears as a line. Try it with the beer mat!

Plate 7.   *Wing movements of the fly* Calliphora erythrocephala L. *An original series of pictures, as seen from above, drawn from frames of a 16-mm cine film. 1–7, downstroke (beginning of stroke omitted); 7–10, reversal at bottom of stroke; 10–20, upstroke; 20–1, reversal at top of stroke. (Right wing is moving slightly ahead of left, indicating that a turning movement is taking place.) 22–4, start of downstroke. Exposure time for each picture about 1/3300 s. This strip of film was taken shortly after the fly started flying, and the legs have not yet been folded into the flying attitude.*

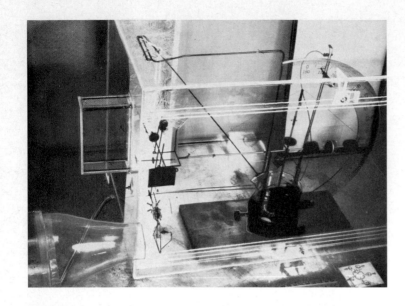
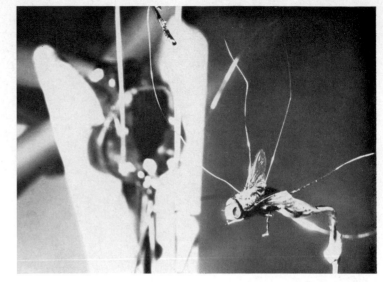

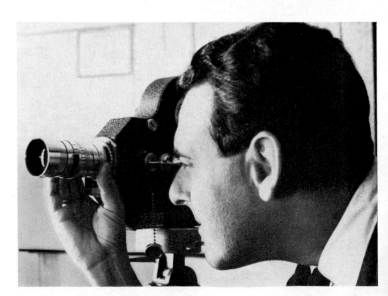

As soon as it performs rotational motions, the line starts tilting backwards and forwards. So that we know which edge is the front, we mark it by sticking on to it a piece of chewing-gum. Now we ask someone to move the mat in front of our eyes in a path which bears some resemblance to the path of the wing tip, and at the same time to rotate it backwards and forwards slowly about its axis. With a little concentration this is quite successful. Now we look at it from some distance away. What we see has been drawn for the insect wing in Fig. 23. Now we know what the lines are, they represent the cross-section of the wing (just as we saw the beer mat in cross-section). The small triangle is analogous to the chewing-gum, and marks the leading edge of the wing. On the downstroke it is black, and on the upstroke white. If you look at the curve point for point, you will see that because of the rotational movements the angle that the cross-section makes with the path of the wing tip is changing continuously. We say that it is altering the angle of pitch. It is, to mention it in passing, the first indication that we have of the complicated functioning of the basal hinge, which has to impart its intricate movements to the wing. On the downstroke, the angle is large to start with, and then decreases, reaching a minimum at the middle of the stroke. At this point the lines lie nearly flat on the wing tip path like the scales of a fish. Then the angle increases again. On the upstroke the angle changes its size and sign in a rather complicated way.

If we confine our attention to the important wide surface of the outer two-thirds of the wing, we can summarise all this by saying that the motion of the wing is the result of a complicated interplay between a beating and a rotational movement, which is repeated exactly in each complete up and down movement. By plotting the path of one point on the long axis of the wing, for example the wing tip, and the rotation of the wing about the axis, the whole complex motion is almost completely described.

There is, however, a third variable, the torsion. Take a

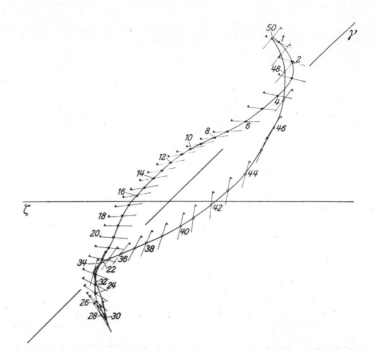

Fig. 23. An unrolled or flattened out wing tip path during one complete up and down movement. The fly is oriented as in Fig. 26. The interval between the points is 1/6400 s. At each point the plane of the wing—represented by a line—is drawn in. The small traingle is on the leading edge of the wing, and it sits on the upper surface. On the downstroke it is filled in and on the upstroke it is left hollow. The axis ζ corresponds to the longitudinal axis of the fly (median), and the axis γ is the best fit of a straight line to the path of the wing; it lies at an angle of about 45° to the median. The projection onto a flat surface has been made without applying any corrections, and at the ends of the path there are noticeable distortions.

piece of paper between the right and the left hands, and then rotate your hands relative to one another. The piece of paper is twisted first in one direction and then in the other. We have described the wing movements under the assumption that the wings do not twist. This is in fact only the case for the outer two-thirds of the wing, which is the important part aerodynamically. On the other hand, the inner third of the wing twists strongly. However, we don't wish to discuss this third type of movement in detail, because the inner third of the wing is not very significant aerodynamically. If you remember that the fly is rushing about the room, you are certain to think again of the translational movement which has already been discussed. Now we have gathered together all the four types of movement which occur during insect flight:

1. Beating movements
2. Rotational movements
3. Twisting movements
4. Translational movements.

Point 3 can be largely neglected, and point 4 will not interest us until we come to consider the aerodynamic forces. Points 1 and 2 remain for our short 'history of a two-hundredth of a second'. For this purpose look at Fig. 24 and Plate 7 (p. 43) in conjunction with Fig. 23.

Let us begin with the downstroke of the wing. The two wings are fully raised, with the wing tips almost meeting over the abdomen. After pausing for a few ten-thousandths of a second they begin to move obliquely, forwards and downwards, at first slowly, and then faster and faster. At the same time they twist sharply, and the angle of attack, which was originally an excessive one of more than 90°, is suddenly reduced. The powerful flight muscles of the wings contract more and more quickly until about the middle of the downstroke, at which point the wings have reached their maximum speed and the angle of attack is at its minimum. From then on the wing slows down and begins to rotate in the opposite direction, with the angle of attack increasing again. The downward beat of the wings becomes more oblique, and the speed falls quickly to zero at the bottom of the stroke.

The wing again pauses for a few ten-thousandths of a second, but the twisting continues, and is in fact at its greatest at the bottom of the stroke. The wings twist further and further round their longitudinal axis until eventually the surface that was uppermost during the downstroke is now facing downwards. In Fig. 24 this surface is stippled, and the diagram shows how more and more of it is exposed.

Before this rotation is complete the whole wing is suddenly jerked upwards and obliquely backwards. The twisting is confined to the inner third of the axis of the wing, while the outer two-thirds remains flat, as shown in Plate 7 on p. 43. The upstroke continues at an angle of 45° and so the wing traces out a path that lies behind the path of the downstroke. The direction of rotation now reverses again, and the angle of attack increases, but just as it becomes dangerously great the wing reaches the top of its stroke and pauses before beginning the next downstroke.

During the incredibly short time of one complete stroke the wings have passed from their initial big angle of attack through a complete inversion and back again. The wing has changed direction twice, and its rotation has reversed no less than four times, once in the middle of the downstroke, once in the middle of the upstroke, once again towards the end of the upstroke, and finally at the top of the stroke. These changes take place in every stroke, and at these precise points and nowhere else. Although the frequency of the wing beat can be varied they always twist with twice the frequency with which they beat, and the oblique and rotational movements are strongly synchronised, and bear a definite phase relationship to one another. This cycle is repeated nearly three-quarters of a million times per hour, and a 'good' fly

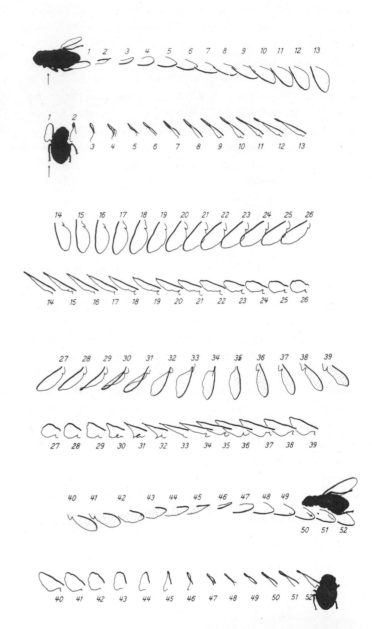

Fig. 24. *The position of the wing of the fly* Phormia regina *during flight in the wind tunnel, projected onto the stroke plane (upper row), that is a plane at 45° to the long axis of the body. The interval between the drawings is 0.17 ms, or 1/6000 sec. The leading edge of the wing is drawn with a heavier line, and the underside of the wing, as far as it is visible, is stippled. The animal was flying steadily at 2.20 m/s in front of the wind tunnel, and the forces were balanced as is shown in Fig. 17. The pictures are drawn in such a way that the points indicated by the arrows (No. 1!) lie directly under one another. The rotation of the wing at the beginning of the upstroke can be seen clearly (starting at No. 27); the morphological underside comes more and more into view, and remains the physiological upper side for the whole of the upstroke. [Note added by translator: the lower row of drawings shows the wing projected onto a plane at right angles to the stroke plane showing the appearance of the wing from above and behind the animal.]*

can keep this up in a wind tunnel for two or three hours without interruption.

What is the function of the twisting of the wing, and how far can this be inferred from the photographs and drawings? It seems likely that the function of the rotation is to ensure that the leading edge always leads, by pointing in the direction in which the wing is moving at any particular time. We have seen in Chapter 3 that the front edge of the wing is thickened, stiffened with veins, and furnished with rows of bristles. If the wing did not rotate so that this stiff leading edge pointed obliquely downwards at the top of the stroke and obliquely upwards at the bottom then all this beautiful construction would be wasted. During the upstroke the soft flexible hind margin would go first, would bend and curl under the wind pressure, and would be a source of wind resistance. Thus this whole contrivance has the single function of avoiding undesirable wind resistance in the most ingenious and economical way possible.

We shall say more about wind pressure in the next chapter.

15. How the wing beats generate aerodynamic forces

Today, in the era of the jet plane, propeller aircraft are already almost obsolete. Giant jets cross oceans and great continents in a few hours, and if supersonic aircraft come into operation in the seventies, the flight time across the Atlantic will be cut to only two hours.

In the living world, jet propulstion is rare and takes place only in water among such creatures as dragonfly larvae, molluscs and squids. Supersonic flight is unknown, though at one time it was erroneously reported in a North American deer botfly. This was often mentioned in the popular press as a 'Wonder of Nature', but it has long been discredited. Simple energy calculations of the kind that we shall hear about in Chapter 25 at once show the impossibility of such speeds in insects.

The wing beats of insects, bats and birds have some mechanical similarities to the old-fashioned propeller, but with the important limitation that wings cannot revolve like an airscrew, and can only beat to and fro with a small amount of rotation. Revolution is both easier and more reliable than reciprocating motion for a machine, especially when heavy masses have to be moved. Nature, on the other hand, knows nothing of revolution round a shaft, and the potentialities, structural materials and size range of natural organisms all make the beating of wings the only feasible flight mechanism for animals.

There is another big difference between the two systems. In an airliner the propellers are called upon to produce only the forward traction, or thrust, and the fixed wing supplies the lift, whereas the wings of insects must serve both functions. A fly flying straight ahead resembles a helicopter more than a fixed-wing aircraft. A helicopter in 'level' flight is really flying obliquely, with its back end tilted upwards in such a way that the plane of its rotors is inclined to the plane of flight. The rotor blades move forwards and downwards, upwards and backwards, just like the wings of the insect. If the helicopter pilot wishes to ascend vertically he brings the rotor exactly into the horizontal plane, so that it draws air from above and flings it away directly downwards. The effect of this is to subject the helicopter itself to a reaction force which is directed upwards and which causes the helicopter to climb. In horizontal flight things are slightly different. Imagine the helicopter to be flying from right to left. The rotor then draws air in and flings it away obliquely downwards and to our right, so that the relative wind past the helicopter is blowing obliquely downwards and backwards. The reaction force on the helicopter is in exactly the opposite direction, forwards and upwards. This is our old friend the resultant aerodynamic force, which can be resolved into two components at right angles, a lift component L which supports the helicopter in the air, and a thrust component T which propels it forwards. The helicopter is able to improve its forward thrust by making periodic variations of pitch, and by other devices.

Fig. 17 (p. 35) shows that exactly the same happens in the fly. As soon as the components L and T are big enough to balance W and D, the forces are in balanced pairs, and the fly is in equilibrium in steady forward flight, a situation that we have already discussed at length. This then is the aim and purpose of applying power by oblique strokes, both in nature and mechanically. The air must always be propelled backwards and downwards, so that the reaction force will provide the two desired components, L and T. This oblique driving force is present in all single-rotor helicopters, and in all insects, bats and birds, without exception.

Plate 16 on   shows a bluebottle as it flies in the stream from a wind tunnel. A narrow beam of very bright light illuminates a system of parallel smoke trails, which stream out of the wind tunnel and make the flow of air visible. The camera is focussed on the left wing (bottom right), and the system of force balances is not seen because it lies behind the plane of the focus and is not illuminated. Note how those horizontal smoke streams (from the left) that are picked up by the moving wing mingle together and appear as a milk-white band running backwards and downwards (to the right). This photograph exactly matches the diagram of forces shown in Fig. 25. For comparison, the bottom left-hand photograph shows the smoke streams beside a wing that is not moving. There is no trace of the band running backwards and downwards, which must therefore be attributed to the wing movement.

The first photograph of Plate 16 (bottom right) suggests that changes in direction of this oblique air stream can be used as a means of control during flight. Suppose that until now the insect has been flying straight ahead and that then it suddenly changes the angle of the emergent air stream to a steeper one. Driving the air more downwards and less backwards will cause the reaction force to be directed more steeply upwards, with the effect of giving a bigger lift component and a smaller thrust component. As a result the insect, which previously flew at a steady height above the ground, will suddenly swoop upwards in a rising curve.

Thus we can see that in the insect world there are a whole range of possibilities for flight control, some of them quite delicate. One insect will prefer one method, another insect a different one. Our cine films have shown that the complicated rotation of the wings serves to vary the angle of attack. In the polar diagrams of the scarce swallowtail butterfly and other insects (Chapter 8) we have seen how important is the angle of attack in producing the right aerodynamic forces, and we have now learned that the end result of the whole wing beat is to deflect the air mass and accelerate it, and that the desired components of force (lift and thrust) arise directly from this. During the course of each stroke the rotation of the wing varies the angle of attack in a regular fashion, and thereby varies the speed and direction of the air that is seized and driven away by the wings. The resultant reaction force is continuously resolved into the components of lift and thrust, the

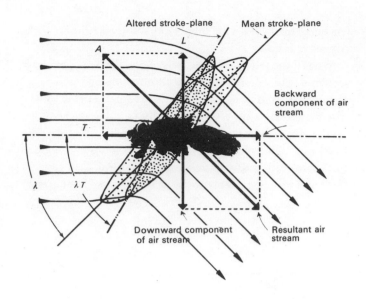

Fig. 25. Diagrammatic representation of the airflow through the 'propeller' of a fly in free flight. The mean stroke-plane (lightly stippled oval) makes an angle $\lambda = 38°$ with the longitudinal axis of the insect. The movement of the wings brings about a constriction of the air stream, thereby accelerating it, and at the same time deflects it obliquely downwards. The equal and opposite reaction on the wings takes the form of a force directed obliquely upwards and forwards. This force A can be resolved into two components at right angles, the lift L directed upwards and the thrust T directed forwards. The fly can vary the direction of the stroke-plane of its wings, which may become, for example $T = 48°$ (densely stippled oval), while at the same time reducing the amplitude of wing beat. Because of the smaller amplitude of the wing beat, the force A is reduced, but because of the greater angle of the stroke-plane more of this force goes into thrust and less into lift.

values of which therefore change in a regular sequence during the course of a single wing beat. The whole complicated cycle is repeated during each stroke of the wings, and is taking place a thousandfold every few seconds.

If, however, we sum up all these variations during the first stroke of the wing; if we take a mean, so to speak, of all the points, numbered 1–50, in the curve in Fig. 23, then the picture changes. In mathematical terms we integrate the component forces over the period of one wing stroke. From these many individual values we arrive at just two final figures, the mean thrust and the mean lift. If we carry out the same operation for the second stroke we shall arrive at the same two values, and again for the next stroke, and the next, indeed for every wing stroke so long as the insect maintains a steady flight.

These values no longer correspond to the components L and T as shown in Fig. 17. Up until now we have tacitly assumed a 'static' condition in which each of these components maintained a constant value, yet, as we have just seen, each component is in fact subject to regular variation during every stroke of the wing. If the simple, 'static' diagram of Fig. 17 is looked at again 'dynamically', the mind's eye can picture the actual movement. Imagine the wing passing through a single stroke, from top null point to bottom, and back again. The two arrows representing lift and thrust begin to change rhythmically, becoming now longer, now shorter, than they are shown in the diagram, according to the point reached in the cycle and the angle of attack of the wing. The direction of the arrows, of course, always remains the same. Now repeat the operation faster and faster, until the changes can no longer be detected. Although the two arrows are still oscillating rhythmically they seem, to the sluggish eye, to take up fixed lengths and to keep to them. These are, in fact, the lengths of the arrows as they are drawn, and they are the results of the integration.

Figure 17 should not therefore be thought of as a kind of instantaneous photograph, as if a flash exposure had 'frozen' one particular wing beat. On the contrary it must be seen as a time exposure in which the details are blurred but which still gives an indication of both magnitude and direction. The experiment with the smoke trails shown in Plate 6 is relevant

here because the time of exposure was one second, and so covered more than two hundred wing beats. No matter how narrow the beams that are cut through, no individual smoke trails can be seen.

So the arrows shown in Fig. 17 must be thought of as fluctuating in length, yet the values of W and D do not fluctuate: the weight of the fly (W) remains constant, and so does the drag (D), as long as the fly maintains a constant velocity. This looks like an inconsistency. If L fluctuates, but W does not, then the fly ought to rise and fall rhythmically. Similarly, if T fluctuates but D remains constant, the fly should move now faster, now more slowly, in the same rhythm as the wing beat. A steady forward speed would seem to be impossible.

This conclusion is correct as far as it goes, but it has never yet been possible to detect such fluctuations along either the horizontal or the vertical axis. The answer must be that the inertia of the fly's body smooths them out, just as the huge mass of the fly-wheel of a steam engine smooths out the rhythmical fluctuations in the forces applied to it by the pistons. Of course the mass of a fly is very small, but it is huge in comparison with the tiny variations in the aerodynamic forces that arise during the wing beat.

16. Glittering wings flit through the room

Finally we must turn to the fourth type of movement, translational movement. We saw in the previous chapter that the beating and rotational movements sufficed for a description of the kinematics but not of the aerodynamics. Here we must also take into account translational motion, i.e.

the movement of the whole flying body from one point in space to another. Instead of the insect co-ordinate system which we have used until now, we must use a fixed co-ordinate system. This in fact sounds more complicated than it really is.

We are concerned once more with the path of the wing tip as shown in Fig. 23. To translate this into a fixed co-ordinate system (see Fig. 26), we add the translational component of the motion to each point. During the time that the wing has moved from position 1 to position 50 and back again, the centre of gravity of the insect has moved a certain distance in a straight line in the direction of flight. For example, if the wind tunnel speed is 200 cm/s and the wings are beating steadily 200 times a second, then this distance is 1 cm (200/200 = 1). We had 50 pictures per wing beat, so the distance travelled between two pictures is 1 cm divided by 50, or 0.2 mm. So to each point in Fig. 23 we add an additional distance of 0.2 mm to the left (the insect is flying in this direction). Point 1 is moved 0.2 mm to the left, point 2 is 0.4 mm, point 3 is 0.6 mm and so on. The whole curve is expanded outwards to the left. This point-for-point construction of the curve is shown in Fig. 27. In the right-hand section, the curve in an insect co-ordinate system is drawn again, and to the left of it the result of transforming this curve to a fixed co-ordinate system. The cross-section of the wing is also shown on the new curve. Its angle relative to the flight path is called the aerodynamic angle of attack.

So at last we have arrived at the curve describing the path of the wing tip through space. We can forget all about the insect co-ordinate system, and look more closely at the new curve. We established at the outset that it must be some kind of hill-and-valley curve. And that is just what we have found. It runs obliquely downwards and obliquely upwards, and

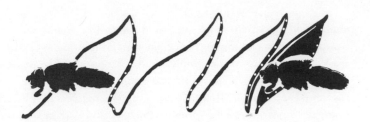

Fig. 26. *The thick black line indicates the path of the wing tip as projected on to a flat surface. Moving from right to left, the path is shown at first as it appears relative to the body of the insect, and this is gradually changed over to show the path through the air. If the insect on the right is assumed to be hovering 'on the spot' in front of the mouth of the wind tunnel, then the wing tip will trace the outline of the solid black shape shown ('insect-related system'). If the wind tunnel is now switched off the wind speed falls off rapidly, let us say within four strokes of the wing down to zero, and the fly moves forward into the mouth of the wind tunnel with increasing speed. From the viewpoint of an observer (i.e. not relative to the fly) the wing tip now traces a series of peaks and troughs that are increasingly farther part ('space-related system'). By the fourth stroke of the wings after the wind tunnel has been switched off the fly has reached the position of the left-hand figure. The flies are drawn from a silhouette cine-film of a blowfly (Calliphora) with the legs cut off. The upstrokes are indicated by white dots enclosed in the black line.*

then a new downstroke begins and the whole process is repeated (part of the next downstroke is shown in Fig. 27, points 52 to 56). Comparing the steadily progressing oscillatory curves of Fig. 21 with the new curve of the path of the wing tip through space, we see that in the latter case the wing

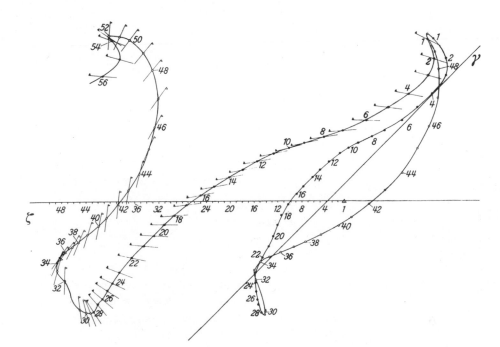

Fig. 27. *The information contained in Fig. 23 is here converted into a space-related system, and the insect-related system is repeated on the right for comparison. The fly is orientated as in Fig. 26 and, as before, is imagined to fly from right to left. The wing tip is shown as a series of short lines, and the angle that these make with the path through the air (the continuous curve) is the angle of attack α at that point. It should be noted that during the upstroke the path through space of the wing itself is actually backwards (points 34–47). Note also the change in the angle of attack during the upstroke, an effect produced by the turning movements (rotation) of the wing at this stage. Each unit along the horizontal axis indicates the distance that a point on the thorax of the fly moves forward between one position of the wing and the next (i.e. in 1/6400 second). Compare also the caption to Fig. 23.*

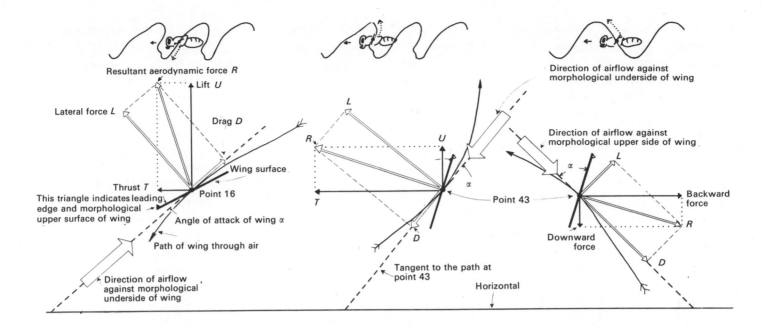

Labels in the figure:

Left diagram:
Resultant aerodynamic force *R*
Lift *U*
Lateral force *L*
Drag *D*
Wing surface
Thrust *T*
Point 16
This triangle indicates leading edge and morphological upper surface of wing
Angle of attack of wing α
Path of wing through air
Direction of airflow against morphological underside of wing

Middle diagram:
L
R
U
α
T
Point 43
D
Tangent to the path at point 43
Horizontal

Right diagram:
Direction of airflow against morphological underside of wing
Direction of airflow against morphological upper side of wing
α
L
Backward force
R
Downward force
D

Fig. 28. A schematic diagram to illustrate how a beating insect wing generates the components of the aerodynamic force. Left: The middle of the downstroke, corresponding to point 16 of Fig. 27. Middle: the first half of the upstroke, corresponding to point 43. Right: Point 43 again, but under the assumption that the wing tip is moving in a sinusoidal path. The orientation of the wing is the same as in the middle drawing. As in Fig. 23, the upper surface of the wing and its leading edge are marked by a small triangle. The sketches above the diagrams show which part of the wing beat is concerned; the small arrows show the direction of flight, the longer and pointed arrows show the direction of motion of the wings. Further explanation is in the text.

tip is travelling backwards during the upstroke! Fig. 27 shows this quite clearly. The insect is flying from right to left. During the downstroke (points 1 to 32) the curve goes from top right to bottom left, as expected. However, during the upstroke it does not go still further to the left, but upwards and backwards to the right until point 46 is reached. Only then does it begin moving to the left again. This unusual path is a result of the fact that in the insect co-ordinate system the path of the upstroke lies to the rear of the path of the downstroke, and we have already seen that this can only be observed in careful experiments using a wind tunnel and force compensation. Under such exactly controlled conditions the normal flight of the insect is faithfully reproduced, and the backwards part of the upstroke can be observed, though carelessly devised experiments will not show this. We can assume that there must be a physical reason for this backward portion of the wing beat in normal flight, and begin to look for it by reconsidering events in the same order as before, starting with the downstroke.

Our hair would turn grey before we had finished, if we attempted to resolve the forces for each and every point on the curve, so we shall restrict ourselves to two examples. We shall choose point 16 on the downstroke and point 43 on the upstroke (Fig. 28), points which are approximately at the mid-points of the strokes. The wings are stretched out away from one another to the maximum extent at these points, and so we can with some justification ignore the sideways components of the forces.

As in the case of the gliding butterfly, the path of the wing through space gives us the direction of the airflow, or more precisely the tangent of the curve at the point in question

gives it. Exactly as before, there is a drag component in the direction of the tangent, and a lift component at right angles to it. We add together the two components geometrically by a parallelogram of forces, and this again gives us the resultant aerodynamic force. So far this is nothing new; we followed the same steps in considering gliding flight, though we can see from the left-hand section of Fig. 28 that the aerodynamic resultant is no longer directed vertically upwards, but is tilted forwards a little. We can no longer call it the upwards force, so we must work out the value of the upwards force by constructing a new parallelogram of forces. The resultant aerodynamic force is simply split into a vertical component *U* and a horizontal thrust component *T*.

We have constructed the resultant aerodynamic force from two components, and then split it up again into two different components. Why have we followed this round-about procedure? We wanted as an end result to know the lift and thrust, forces which act vertically upwards and exactly horizontally, whereas the actual forces acting on the wing were not in these directions. We had to convert them to *U* and *T*, and the easiest way to do this is to first of all to combine them to give a resultant, and then to split up the resultant into two new components, this time in the directions which we require.

So what have we discovered as a result? At point 16 the wing is moving forwards and downwards, and it drives the insect forwards (T) and provides an upwards force (U), just what we would require of an orderly flight system. The situation is similar at all the other points on the downstroke, but the ratio of U to T changes from point to point. The details of these changes do not concern us at the moment.

Now let us turn to point 43 on the upstroke. The wing is moving 'head downwards', as the position of the small triangle shows, in an upwards and backwards direction. The angle of attack α is large. Once again we have the lift and drag components L and D, which are added to give the resultant aerodynamic force. The force is split up into the components U and T, just as for point 16. The result is that once more the insect is simultaneously lifted (U) and driven forwards (T). Once again force components are generated which we would like to see in a good flapping wing flight system (see the centre part of Fig. 28). We should expect this during the upstroke in the same way that we expect it at other times. What would happen if the backwards part of the stroke did not take place, but everything else remained unchanged? The answer is shown in the right-hand part of Fig. 28. We are assuming that the curve would not be as smooth as the central one of Fig. 21. The airflow would impinge on the morphological underside of the wing with a very high angle of attack. The large angle implies a very high drag component, and the resultant force would be directed backwards and downwards, driving the insect backwards and downwards instead of forwards and upwards! The insect could not lift itself, but would push itself downwards. It would not drive itself forwards, but backwards. On the upstroke hindering forces would be generated, nullifying the useful forces produced during the downstroke. The insect would not be able to fly.

So the backwards part of the upstroke has a very definite purpose. It alters the airflow over the wing so completely that it turns the undesirable backward and downward forces into highly desirable lift and thrust! This means that the complicated rotation of the wing only serves a useful purpose when it is combined with an alteration of the path of of the wing. The upstroke must, in the insect co-ordinate system, take place behind the downstroke. As a result of this the backwards section appears in the fixed co-ordinate system, and only as a result of the backwards motion can the angle of attack be chosen to give useful forwards and upwards force components. In turn, all this is possible only because the wing is constructed in such a refined way.

In the case of an arched aircraft wing, the airflow must impinge with only a small angle on the 'morphological' underside. Under these conditions lift is generated. Airflow impinging on the upper surface has catastrophic results. On the other hand, for the flat beating wing of the insect, the flow is on to the morphological underside during the downstroke and on to the morphological upper surface during the upstroke, but in both cases it is able to generate upward and forward forces. The construction of the wing, its changing path and its rotations form a functional entity. Without the different wing-tip paths the rotation would be quite useless, and both would be without purpose if the wing could not function with the airflow directed against either of its surfaces.

This wonderful system is not, however, quite perfect. Even during normal flight, in the second half of the upstroke, small negative forces are generated as the result of unfavourable angles of attack. These unavoidable flaws have disappeared by the time the upper turning point is reached since the immensely fast rotation returns the angle of attack to a 'sensible' value. A tiny snag, but it seems that there is no way of avoiding it. The insect must somehow return the wing to the configuration required for the downstroke, in spite of the fact that there are physical reasons why lift and thrust cannot be generated whilst this is being accomplished. There is no magic in nature. The bluebottle has to fly according to the laws of physics.

I wished to lead you by a series of steps to the following realisation: that the laws of nature do not apply only to the falling ball and the swinging pendulum, but equally to the most complicated mechanisms of the living world. I wanted to make a further point as well: nature is a shrewd inventor. She has to abide by the laws of physics but her constructions are so finely executed and the interplay between the various elements is so minutely worked out, that she has achieved most of the things which are theoretically possible. The technical refinements of nature are of the highest order, and technologists should study her machines. They might learn something!

Plate 9. A big dragonfly, certainly one of the genus Aeschna, *sweeps in to a landing. The legs are outstretched forwards and downwards, and the wings are held with a slight negative dihedral. Taken with a flash exposure of about 1/40,000 s*

Several times already we have noted that in science it often makes a great difference which way we look at a problem. All science is only a way of looking at nature, and we are perfectly entitled to look at it in an intuitive poetical way, but we ought not to call this approach 'unscientific'. Quite the contrary. To lift science out of the commonplace we should employ both viewpoints, logical analysis on the one hand and intuitive sympathetic observation on the other.

Let us look, for a change, at what poets have had to say about insect flight, not forgetting that some of the most precise research programmes have been stimulated in the first place by a chance observation, and some intuitive comparisons. We must not be surprised therefore if experimental biology occasionally betrays us into a lyrical style and a poetical vocabulary.

Friedrich Zschokke, for fifty years Professor of Zoology at Basle, wrote enthusiastically about the intricacies of flight:

'Like a whirling snowstorm, myriads of mayflies swarm and soar over the water which gave birth to them, which sheltered their eggs, and which several hours later will be their tomb. Male midges dance in the golden sunshine of a fading June day, their bodies flashing in the streamers of light, and then vanishing into the shadows. Great stagbeetles make heavy lumbering flights high above the tops of oak trees. A robberfly hangs trembling in the air like a hawk, ready at any instant to pounce on its prey, and a hoverfly flits nervously from blossom to blossom on its gossamer wings. Bombyliidae, beeflies, disport themselves on sunny hillsides, over the meadow flowers, hovering for for a brief second in front of a fragrant blossom, their wings whirring in the air like a humming bird, and their

long proboscis sunk deep into the nectar while they are still in flight. Then off they go like a flash to feed somewhere else, so quickly that the eye can hardly follow them.

Among the most accomplished fliers, combining grace with agility, are the dragonflies, especially the delicate damselflies. Long ago someone called dragonflies 'shining Amazons', and marvelled at their thousand different motions, their twisting and turning, backwards and forwards, in never-ending circuits, now over a meadow, now over the reedy margin of a pond or stream.

But it is the butterflies, like aerial blossoms, that really show the possibilities of flight. Moths fly from house to house, straight for the nearest open lighted window, to the nightly alarm of the housewife. While the landscape is still half covered with snow the brimstone butterfly comes staggering out of its winter sleep, like a harbinger of spring. Butterflies such as the camberwell beauty and the peacock, flit about the blooms in awkward flight on their two pairs of broad equally-sized wings. They deserve the name "sommervögel" which they have been given in poetry and folklore.

Night-flying moths fly more purposefully, with elongate narrow forewings and much smaller hind wings. The privet hawk moth is like a flying arrow in the dusk, pausing for only an instant to drink nectar from a flower, a sudden halt in its swift passage, then away it goes again, replete, into the night. The magnificent apollo butterfly named after the sun god, flutters heavily across the alpine meadows, while the scarce swallowtail glides effortlessly with motionless wings around the sunwarmed rock faces of the Jura.'

(Compare this with the diagram in Fig. 29.) (Taken from *Der Flug der Tiere,* Berlin, Springer-Verlag, 1910, pp. 19–20.)

Let us listen to Jean Henri Fabre, the grand master of insect observation in the south of France. He lived and watched insects in Sevignan, in the Vaucluse through many decades, between 1827 and 1898. The following extract is taken from a delightful account of the nuptial flight of the giant peacock moth *(Saturnia pyri).*

'It has been a memorable evening; I shall call it the evening of the giant peacock. Who does not know this gorgeous moth, one of the biggest of the European night-flying

Plate 10. A skilful hoverer belonging to the genus Bombylius, *she hangs motionless in the air for a fraction of a second. The big picture at the bottom shows the fly hovering over the dark entrance hole of the burrow of a solitary bee, into which she projects her sandy-coloured egg. Note the flying attitude of the bee, particularly the hind legs held backwards and upwards. That this is a stereotyped attitude can be seen by comparing the bottom pictures with the others; the bottom picture is not just another enlargement from the same series as the others, but was taken by a different author in a different country, several years before the others.*

moths, chestnut brown with a collar of white scales? Each of the wings is patterned in grey and brown, with zigzag grey crossbands and edges of the same colour, and in the middle is a huge eye-spot with black eyeball and an iris of shifting colours in which black, white, chestnut-brown and amaranthine rings run into one another.

On the evening of 6 May a female of this moth emerged from its pupa on the table of my laboratory. I put it into a wire cage while it was still soft, with no other idea than to see what it would do next. About nine o'clock a great disturbance began in the room next to mine, and my little boy Paul, who was still only half undressed, started jumping and dashing about in a crazy way, knocking over the furniture and calling out "Come quickly, come and see some moths that are as big as birds. They are filling the whole room". I hurried in, and saw that the boy's excitement was fully justified. Our house was being invaded by giant moths. Four had already come in, and he had put them into a bird-cage, while many others were beating at the window. I suddenly remembered the female that I had shut up until next morning. "Get dressed again," I said to my son, "and come with me. We are about to see something worth watching." On the way to my study, which was in the right wing of the building, I passed through the kitchen, and asked the maid if she had ever seen moths so big that they could be mistaken for bats at first sight. The uproar seemed to have affected everybody in the house, but what had happened to the original cause of the disturbance? It so happened that one of the two windows in the room had been left open, so that the way in toward her cage was clear.

Candle in hand, we crept into the room, and what we saw there is still unforgettable. Huge moths flew round the wire cage, lazily flapping their wings, settling, going away, coming back, flying up to the ceiling, then down again. They dived at the candle, and blew it out with the draught of their wings. They settled on our shoulders, beat at our clothes, and struck us in the face. This whirl of moths had something ghostly about it, and little Paul gripped my hand more tightly than ever.'

(From *Bilder aus der Insektwelt,* Stuttgart, Franckh. 1908, pp. 80–1.)

The same author described in another place how hunting wasps bring their prey down to the ground. In this instance the prey was a big bumblebee, overpowered and killed by the wasp *Bembex tarsata.* Fabre wrote:

'The nest lies at the foot of a steep, sandy slope. The wasp announces its arrival with a loud hum, which has a mournful note, and which does not cease until the insect has alighted on the ground. The wasp circles over the slope, then descends vertically, slowly and cautiously, but still making

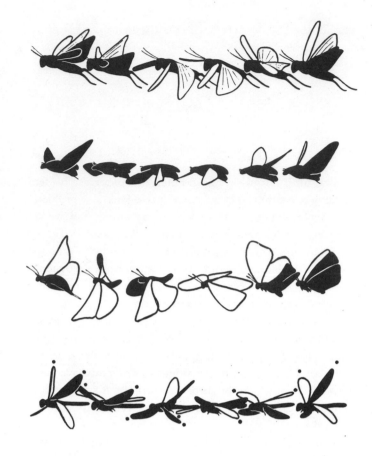

Fig. 29. Sketches of the wing movement of several different types of insect, drawn with the aid of time-lapse photography, and always reading from right to left. From above downwards: locust, beetle, butterfly, dragonfly. The upper surface of the wing is shown in white and the lower surface in black. The dragonfly (bottom row) beats fore and hind wings in exactly opposite phase: when the fore wings (marked with a black spot above the tip) beat downwards, the hind wings rise up. The locust (top row) shows this phase difference much less pronounced. The butterfly, because of its wing-coupling mechanism, beats its two wings exactly in phase.

the same clear hum. If its penetrating eye sees anything unusual it slows its descent, circles for an instant and then climbs up again, turns away and is off like lightning. A few minutes later it is back again, and hanging there at a particular height, as though watching from a lookout. The cautious descent begins again. Finally the hovering wasp is reassured about a patch of ground which, to my eye at least, looks in no way different from the rest. At the moment of alighting the plaintive note is cut off.'

Elsewhere Fabre described what the hunting wasp does if a parasitic fly, belonging to the genus *Miltogramma,* is waiting for it on the ground, when the following very interesting situation occurs. Normally, the wasp lays an egg on its prey and then buries it in the sand; the wasp larva, when it hatches, feeds upon the prey. When the larva is fully fed it pupates, and from the pupa emerges another winged adult wasp which catches fresh prey and stings it. This is the life-cycle of this kind of wasp.

The parasitic fly lies in wait for the wasp, and at the right moment lays its own egg on the wasp's prey. The maggot which hatches from the fly's egg develops more quickly than the wasp larva, robbing the latter of most of the food stored for it, and thereby weakening it so much that the wasp larva never manages to pupate. If food is short the wasp larva may even be eaten by the maggot. A wasp flying down with prey is well aware of the danger, but strangely enough makes no attempt even to attack, let alone to kill the fly. Instead, there takes place an aerial game of skill to see which of them is the nimbler. Fabre tells this tale in his inimitable style:

'The flies sit around on leaves in greater or smaller numbers, usually threes or fours, and completely motionless. All are looking towards the ground, and they know in what direction it lies even if it is out of their sight. Their dark brown colour, their big blood-red eyes and their unwavering stillness often gives me the impression of a group of bandits in grey cloaks with red kerchiefs round their heads, waiting for the moment to strike. The wasp arrives with her prey. Even if nothing untoward was to be seen she would still hesitate, circling at a particular height and descending only slowly and cautiously. A shrill note of vibrating wings shows that she has been alarmed by seeing the villains. They, in turn, have seen her, and are following her with their eyes, as their red heads show, all turned towards the coveted prey. Now begins a ding-dong struggle between caution and rashness. The hunting wasp descends slowly with scarcely perceptible motion of its wings, as if she were gliding down and using her wings only as air-brakes. Now she circles in a patch of sunlight; this is the right moment for the flies to move. They rise up and take position behind the wasp, one close behind, the others further off, even flying in formation. If the wasp tries to get out of their sight by turning, they turn too, and with such precision that they end up still behind her in the same formation. Whichever way the wasp flies they do the same, following the lead of the wasp all the time. They make no attempt to reach the prey which is the object of all their manoeuvres. Their tactics are confined to lying low, keeping themselves in the background, reserving themselves for the swift climax that comes at the end of every delaying action.

Sometimes the wasp gets tired of the waiting game, and drops to the ground, whereupon the others at once stop pursuing it, and also settle. The wasp takes off again with loud hum; this is the signal for the flies to follow suit and follow it even more persistently. There is only one way left to try and shake off the pursuers. The wasp flies off in a furious burst of speed, trying to outdistance the flies, but the crafty flies do not fall for this trick. They let the wasp go off on its own, and settle down again on the leaves to watch the ground. If the wasp comes back for another attempt, the whole sequence begins again, until eventually the obstinacy of the parasites wears down the maternal caution of the wasp. The moment the wasp relaxes its vigilance the flies are on the spot. The one that is most conveniently placed pounces on the prey before it disappears from sight, and lays an egg on it.'

(From *Aus der Wunderwelt der Insekten*, Mesenheim, Hain, 1950, pp. 90, 102–3.)

Let us see what two other biologists had to say. Fritz Schremmer wrote about the 'Attempts at concealed egglaying by beeflies'. These are two-winged flies, the larvae of which are parasitic in the nests of ground-living bees. Before they lay their eggs, these small colonial bees dig out a chamber in the sand with the tip of their abdomen, and afterwards camouflage it with finely powdered sand. In spite of the deliberately neutral and factual language of the following account, we can sense the excitement of the author when he was making these observations.

'*Bombylius vulpinus* is a small hairy species, predominantly yellow-brown or grey in colour, which develops as a parasite of the small bee *Panurgus calcaratus*. This bee is covered with shaggy black hairs, and nests in more or less dense colonies in sand or firm loamy soil. The entrance holes appear black to the naked eye, and are four or five millimetres across. The beefly visits the colonies in the morning between nine and eleven o'clock when the bees are most active. Flying with an audible, high-pitched hum two or three centimetres above the ground, and approaching each burrow in turn, it descends to a height of about one centimetre and hovers, poised with its eyes exactly over the opening (Plate 10, p. 54). At the tip of its abdomen the beefly has a double brush of fine, silky hairs which show up rather paler, and which are pressed together in flight. When the fly is hovering over a nest opening it makes two or three, or perhaps more, characteristic flicks with the tip of its abdomen towards the opening. At each flick an egg is projected towards the mouth of the nest.

At this point we might say a little about the characteristic attitude of the beefly in flight. Fore and middle legs are held together, and stretched forwards parallel to the proboscis, while the hind legs are extended backwards with the tarsi stiffly angled upwards. It would seem as if this attitude of the legs might have some importance aerodynamically.'

(*Verhandlungen der Deutschen Zoologischen Gesellschaft* (*München*, 1963). Leipzig, Portig, 1964, pp. 293–4.)

This manoeuvre has rarely been photographed, but Plate 10 shows beautifully the position of the legs. Since the hind legs project downwind in relation to the fly's motion they may well have a steering function. The tiny vinegar fly, *Drosophila*, steers itself in flight on a similar principle.

The Danish biologist Carl Wesenberg-Lund wrote a wonderful book on the biology of insects in fresh water, from which the following short and striking account is taken. It relates an observation on the non-stop flight of a big dragonfly (the libellulids, or true dragonflies, in contrast to the fragile damselflies).

'Like many birds, dragonflies catch their prey in the air, but unlike most birds they do not have to alight before they eat it; they can tear their prey to pieces, chew it, and swallow it while still continuing to fly. Many dragonflies also begin the mating process in the air and some are able to complete it there, while most also lay their eggs from the air. Let us watch a *Libellula quadrimaculata* as she flies over the still surface of a woodland pool, and note how she first catches and devours a victim, then mates with another dragonfly in flight, and immediately afterwards begins to lay eggs. All these vital functions are carried out one after the other in continuous flight, without any need to alight on the ground or on the vegetation in or out of the water. As far as I know the Odonata are the only creatures living today of which this is true. No pen could describe the consummate elegance with which this insect, poised in the air, changes from one operation to the other; the precision with which the prey is pursued and captured; the incredible speed with which the copulatory organs are brought into position, ejaculation takes place, and the pair separate again. There is only one thing they cannot do in flight, and that is rest.'

(*Biologie der Süsswasserinsekten*, Berlin, Springer-Verlag, 1943, pp. 55–6.)

The author of this extract was a level-headed biologist, who avoided talking about the dragonfly as a 'wonder of nature', and confined himself to such terms as 'organism', 'individual' and 'biological functions'. Nevertheless, in every line of his writing we can feel how much he was moved by what he saw.

This gives us a hint that there are things to which only a poet can do justice, and at this point we come to the end of this little homily on enthusiastic observation, and back to the original extracts from Zschokke's works, in particular to the soaring and gliding of the mayflies. These insects often spend a year or more as nymphs burrowing in the mud under water. Every few days a large number of them rise to the surface and emerge simultaneously as a cloud of winged adults, which—at least in some species—moult once more, including the wings, to give the sexually mature adult. They form mating swarms of millions of individuals which pair in the air, glide wearily down to lay their eggs in the water, and then die. Sometimes their dead bodies lie in heaps along the shore line, while the fertilised eggs in the water are the start of a new generation.

Wesenberg-Lund has this to say about them:

'What are all the legends and fables about animals compared with the fantastic truth of nature itself? With our own eyes we can see creatures with their mouthparts, so to speak, withered way; with the skin inflated with air until it functions as an air bladder; with males so small that during the mating flight the females can drag them along; with legs of various rudimentary shapes; with tail bristles modified into an egglaying apparatus; with their eyes elaborated into the most complicated organs of any in the insect world. Moreover, these wonderful insects are formed, and undergo their development, in an environment far removed from that of their adult life, passing the winter as nymphs under the ice in lakes and marshes. When spring comes they press on with their development, and one fine day the finished mayfly lies ready within its nymphal skin.'

For what happens next, let us listen to someone who can speak in poetic language, Franz Graf Zedwitz.

'At this point they emerge on to a reed stem, or projecting root, on to willow trunk or the slope of the bank. Yesterday there were a few of them, today there are hundreds, tomorrow they will be milling about in uncountable hosts. They burst out of their skins and become winged creatures, and look like fully developed mayflies, with four milky-white wings pressed together above the body. Why then do they not fly? Why do they sit about like winged squadrons, waiting to take off into the warmth of the summer afternoon, into the fragrance of the meadow, and the fresh air over the water? What are they waiting for?

The sun sinks, and slants in diagonal shafts across land and water. The evening breeze blows fitfully, hardly stirring the leaves and the meadow grass. And now comes the miracle of the evening and the night. Was there ever a summer as replete as this? Has even this summer seen such an evening? There is a stirring, a rustling, with here and there a fluttering movement up the stems. Skins are shed, the winged insects moulting yet once more, almost, but not quite, in one operation. Out come the antennae, the legs, the long tail bristles, and then, miraculously, the wings, which have blood pulsating through them, and are not stiff and hard like the wings of flies, butterflies and bees. Is this why they have been waiting these two, three, five hours? Their time is come. Their few hours of adult life have begun and the first of them are already flying off into the cool evening air, fragile creatures, with shimmering wings, and big eyes, glistening in the setting sun. How delicate they are! Such beings belong to the land of dreams!

Blood pulses through their wings, the urge for fulfilment possesses them. They do not thirst for the nectar of flowers, but for life itself. In their tens of thousands they wheel and glide with the rustle of elfin wings, building up into a rising, falling, dancing cloud like smoke. The moon rises, making the water look like black lacquer, and the ripples like flashes of sunken treasure. The ghostly cloud of mayflies rises higher and higher, billowing out as far as the eye can see, while within it the big-eyed males and small-eyed females search for and find each other.

The moon hangs in the sky. How much time has passed? Is it really already time to die? Is the dream ended, the illusion over? The hare is still feeding in the field. The great red deer peers suspiciously out of the wood, like a rustling ghost. Now a male mayfly falls, dead from exhaustion. Female after female tumbles down towards the water.

Down they go, falling like snow, covering the dark surface of the water. They let their eggs slip out into the water, give a final quiver and then die.

This then is a tale of creatures of the air, of the life and death of phantoms. O Death, how soon you come, out of the wonderful dream of a summer night.'
(*Wunderbare kleine Welt*. Berlin, Safari, pp. 60–62.)

Emilia Slavuy *Fireflies*

In dusky moonlight	Sparkles a firefly
On filmy wings borne	Meadow to hedgerow
Twinkling	
Dancing	Elegant lantern
On bough and treetrunk	Shines in the twilight
	Flies back and forward
Through the soft evening	Fluttering down
To its love of the gloaming	Blossoms and dewdrops
Windmills its journey	

18. The precision mechanism that drives the wings

Can you imagine what an insect thorax looks like in cross-section? If an Hungarian salami is cut, and then viewed exactly end on, what is seen is a red disc punctuated with bright spots, and nothing more. What lies behind cannot be seen, however long the sausage is (provided that it is straight). One has obtained a two dimensional section of a three dimensional body.

Now let us imagine that a biologist does the same thing with two freshly killed insects, one of them is our old friend the housefly, and the other a large grasshopper. The biologist does not use a pen-knife to make the cut, but a special instrument called a microtome. We ask him to study how the wings are set into the body of the insect and to draw what he sees. At the end of two months he shows us two beautiful drawings: very impressively complicated, but we don't understand anything, even when he explains his work using all the correct technical terms. Finally we ask him to simplify it for us,

to make a schematic drawing. The biologist puts on a thoughtful face, 'I can easily draw the basic principles for you', he says, 'and you will be able to see clearly how the wings are moved, but only—I must repeat—in principle. The drawing will not be a true representation of nature, but much, very much simplified and schematised. Do you mean something like this?' And in a few minutes he has drawn two sketches (Figs. 30 and 32). That is exactly what we do mean. If we don't begin in a simple way we might as well give up at the start, because there are so many complications.

First of all we should agree as to what these diagrams represent. Only one upstroke and one downstroke muscle on each side is shown. Contracted muscles, that is, ones which have shortened, are shown as stippled with large black dots. Muscles which are not contracting are not stippled. The lower and side plates of the thoracic box are shown fused. Joints are represented simply as a small thick-lined circle, and we should remember that they allow the wings to turn in the same way that the rudder of a boat can turn in its bearings. In each case one of the drawings shows the position of the wings during the upstroke, the one in which the wings are higher, and the other shows the position during the downstroke.

Let us consider the locust first as it is easier to understand. In longitudinal section the wing looks like a lever with unequal length arms. If the outer, or downstroke, muscle contracts, the tip of the wing moves downwards, and if the inner, or upstroke, muscle contracts, the tip moves upwards. Very fine foldable membranes cover the 'open' parts of the hinge. This describes completely the principle of the way the wings are moved. We immediately notice two things.

From the point of view of muscle physiology, the two wings can work completely independently. For example, only the muscles on the right-hand side might be working and not those on the left, then the insect would bank steeply and dangerously to the left. The nervous system must give the command 'contract' to the right- and left-hand muscles at exactly the same time, either to both the upstroke muscles or both the downstroke muscles. Only under these conditions will the right and left wings move with the same velocity, and the locust fly in a straight line. On a migration of perhaps 300 kilometres the nerve impulses must arrive simultaneously on many thousand different occasions. This is the first point to notice: the simultaneous stimulation of the right and left halves of the insect.

Another point is equally obvious: the flight system needs to be stimulated in an exact rhythmic sequence. Let us assume that the wings are at their lowest point. Then, and only then, must the central nervous system—the CNS—give the order to the two upstroke muscles, 'upstroke muscles contract to raise the wings', whereupon the two muscles contract and the wings start to move upwards. At the upper turning point the CNS gives the order, 'upstroke muscles relax. At the same time downstroke muscles contract to lower the wings.' So one set of muscles relaxes, and the other contracts, and the wings turn downwards. Once more the CNS orders, 'downstroke muscles relax. At the same time upstroke muscles contract to raise the wing.' The whole process begins over again.

So you see that this flight system needs continuous attention. During each stroke the CNS must cause the muscles to contract at the right instant otherwise the machine would not work. There must never be a false command, such as, 'up and downstroke muscles contract at the same time'. A correct command must never be given at the wrong moment, for example, 'upstroke muscles contract' while the wings are in the process of moving downwards. A continuous complicated switched nervous activity is necessary: a sort of distri-

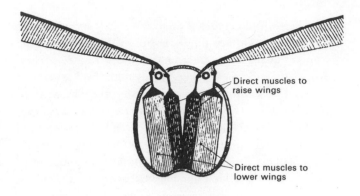

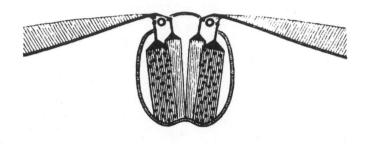

Fig. 30. *Direct flight muscles of the type found in locusts and dragonflies, for example. The hinges are represented by black circles with white centres. Groups of muscles which are contracting are stippled. The upstroke is represented in the upper sketch, and the downstroke in the lower one.*

butor for the flight motor. Because the central nervous system can switch very quickly, and the wing beat frequency is low—in large insects it is 20 to 25 beats per second at the most—the system works well.

But what about insects whose wings beat at ten times this frequency? Suppose the wings are beating 200 times per second, as they do in a small fly? The nervous system literally cannot keep up with it. Nature the inventor must find another way of moving the wings, and the thorax of a fly does indeed work in a completely different way from that of more primitive insects. So refined is the technique used to produce these high frequencies that for once we would be quite justified in calling it a 'miracle'.

In cross-section, the thorax of the fly looks rather like a saucepan with a lid which is a little too small for it. If two wooden spoons are placed between the lid and the rim of the

Strong upstroke muscles run between the upper and lower surfaces of the thoracic box. (Because the upper is called *dorsum* in Latin, and the lower, *venter*, these muscles are also called the dorsoventral muscles.) When these muscles contract, the lid is, as it were, drawn into the pan and the wings move rapidly upwards. The opposing muscles, the downstroke muscles, are also called the longitudinal muscles, because they run the length of the thorax and are joined to the box at each end. So in our diagrams they appear in cross-section like the salami we were talking about. In the drawing (Fig. 32) they lie between the dorsoventral muscles as stippled cross-sections.

In reality there are very many muscles, and they are tightly packed together with no spaces in between. Plate 12 shows this very well. One has to imagine that, as a result of the contraction of the powerful dorsoventral muscles, the whole thorax is distorted and bulges outwards. It is exactly the same as when a tennis ball is squeezed between the thumb and

Fig. 31. A diagram to illustrate the saucepan-lid principle which is used by the indirect flight muscle system of Diptera. (This should be compared with Fig. 34, in which the morphology important to the action of the wing hinge is shown, with the various parts in their correct position relative to one another.)

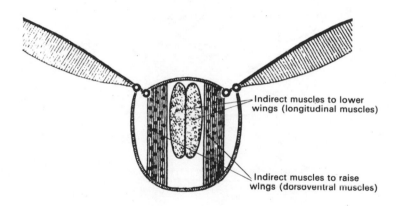

Indirect muscles to lower wings (longitudinal muscles)

Indirect muscles to raise wings (dorsoventral muscles)

pan, one on each side, and the lid is pushed a little way into the saucepan, the other ends of the spoons are forced upwards at high speed. Because of the large ratio of the lever, only a small movement of the lid is necessary. As soon as the lid is lifted a little, the free ends of the spoons move rapidly downwards. Remembering that we are considering a schematised system, we can say that there is a sort of double hinge between the side and the lid of the thoracic box of the fly. To allow the system to work, the plates must not be relatively tightly bound together, as is the case in the locust, but there must be a small split between them, so that they can move relative to one another by a small amount. How is this made possible?

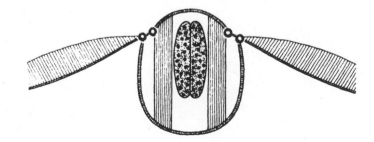

Fig. 32. Indirect flight muscles, of the type found in flies and midges. The conventions are the same as in Fig. 30.

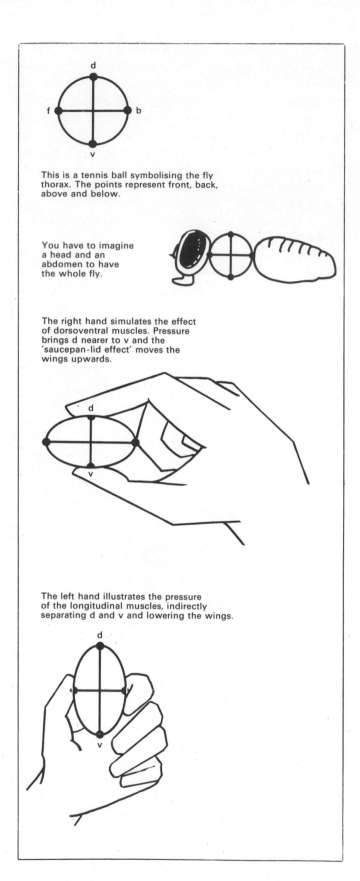

This is a tennis ball symbolising the fly thorax. The points represent front, back, above and below.

You have to imagine a head and an abdomen to have the whole fly.

The right hand simulates the effect of dorsoventral muscles. Pressure brings d nearer to v and the 'saucepan-lid effect' moves the wings upwards.

The left hand illustrates the pressure of the longitudinal muscles, indirectly separating d and v and lowering the wings.

Fig. 33. A diagram to aid the understanding of the deformation of the thorax by the indirect flight muscles of a fly.

forefinger of the right hand: at the top and the bottom the walls approach one another, but all around the sides they are forced apart (Fig. 33). However, as soon as the dorsoventral muscle relaxes and the longitudinal muscle contracts, the deformation is reversed. Not only that, but the longitudinal muscles give the thorax a new deformation in their own direction—they bring the two ends of the box closer together. In this indirect way they force the lid out of the saucepan again, and the wings are moved downwards.

It is astonishing that in the flies, the main flight muscles are not attached to the wings at all. They simply draw the thorax in, front to back and from end to end. In no way do they move the wings directly. Their sole purpose is to cause the thorax to vibrate, so that the small lid on the large saucepan is drawn in and out of the pan by a small amount. As long as this happens, the rest follows automatically. Because of their unique double hinge, the wings are forced to beat up and down with the rhythm of the vibrations of the thorax. This is referred to as the indirect flight musculature of the Diptera, and is in contrast to the situation in the locust, where the muscles act directly on the wings. Locusts and other large insects possess a direct flight musculature. With their sophisticated problem of how to beat their wings at 200 cycles per second, why is it that an indirect musculature is favourable for a high wing-beat frequency? Again there are two basic reasons, one mechanical and the other neurophysiological. Careful consideration of Fig. 31 will demonstrate this.

The mechanical reason can be seen without difficulty. The faster a system is oscillating, the more difficult it is to achieve large amplitudes, because the inertial forces increase with frequency. Because the wings are pivotted very close to their base they act like levers with a large mechanical advantage, and so the amplitude of oscillation of the thoracic box can remain small. The second, neurophysiological, reason calls for a rather deeper investigation.

We have already seen that, in the case of the locust, the central nervous system is able to keep pace with the slow beating of the wings. In the flies this is not possible because the frequency is too high. The first thing that the fly must do

Plate 11. The plume-moth Alucita pentadactyla in a typical resting attitude on the under surface of a leaf. The hind legs are stretched backwards and laid alongside each other, above the abdomen. The feather-like construction of the wings is clearly shown, a central rib fringed with plumes on both sides. The fore wings are divided into two feathers, and the hind wings into three. At rest, the wings are laid alongside each other, but in flight they are expanded like fans.

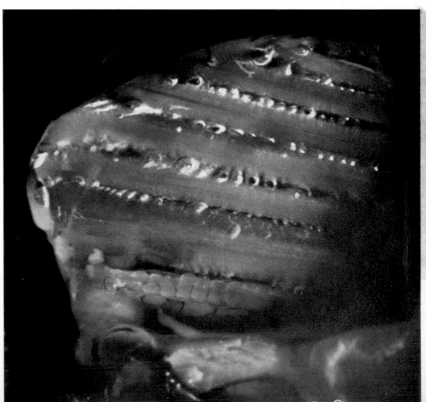

without is the synchronised stimulation of the left- and right-hand set of muscle fibres. The indirect coupling solves this problem. If, for example, the locust sends a nerve impulse to the right-hand upstroke muscle a little later than to the left-hand muscle, the right-hand wing will start to rise later. The insect would cease to fly along a straight path, and might even crash. If the same thing happened to the fly it would not make any difference; all the upstroke muscles, right and left, just pull the lid into the pan and nothing more. If the left-hand muscle is stimulated a little later, it starts to exert tension a little bit later, but this cannot alter the motion of the two wings since they are closely coupled by virtue of the fact that both are moved in the same way by the downwards motion of the lid. A slightly one-sided pull on the lid does not matter because the whole thorax is so stiff.

The fly also has to do without the correct timing of the stimulation of each muscle during the wing beat. It is not possible (in contrast to the situation in the locust) for the central nervous system to deliver its impulse at exactly the right time to muscle No. 1, and then to muscle No. 2 so many milliseconds later, and so on. This still works quite satisfactorily for the more primitive locust with its slow wing-beat frequency, but for the highly developed little fly it is inadequate—a new system is needed. The indirect flight motor is a part of this, but it can only be effective in conjunction with a specially constructed basal hinge and a special type of muscle.

This example shows very clearly a point which we have made already, that Nature does not evolve mechanisms that are complete in themselves, but any particular function can only be served by the co-ordination of several elements. One thing is made to match another, one step carries on where the last left off. The problem was how to make a wing oscillate at very high frequency and the solution is something like this—'By applying *power* through a "snapping disc" *mechanism* in the wing hinge, a particular type of *flight muscle,* and a suitably patterned *stimulation* by the central nervous system.' So there are four items, of which we have already considered the first. We shall consider the sophisticated wing joint in the next chapter, and the flight muscles and the rhythm of the nervous stimulation will be left until later.

19. The wing joint as a tiny click mechanism

Just now I described the wing joints as 'snapping discs'; this can easily be explained. Take an empty shoe-polish tin with a rounded lid, and press slowly on the lid. It bends inward more and more, and then, suddenly, with an audible snap, it

Plate 12. *The flight musculature of a blowfly (Calliphora erythrocephala) dissected out under water so that the associated air sacs can be seen. All three photographs are the same way round, with the head to the left, and the dorsal surface upwards; the two lower pictures are less enlarged than the upper. Removal of the left wall of the body exposes first of all the left dorsoventral muscles (bottom left), and then comes to the beginning of the left dorsal longitudinal muscle (above). The confused network of shining tracheae is seen most clearly at this stage. Further dissection destroys the tracheal loops, but gives a better view of the cross-sections of tracheal trunks (bottom right), some of them expanding into air sacs, which encircle each of the separate longitudinal muscles.*

becomes completely bent in. The direction of the arch has been reversed. If you then—say, through a hole in the bottom —press it slowly outward again, the same thing happens. At a certain position the lid jumps back into its normal outwardly-curved position, snapping again as it does so. There is a sudden abrupt transition from one stable position into a second.

A similar sort of operation is found in two-way light switches, and particularly in the modern microswitches built into electrical devices (Fig. 34). They have two stable positions, in and out, and they snap instantly from one to the other when bent by a certain amount into a critical position, and not before. In technology such systems in general are called bistables, or flip-flops.

The wing joints of the flies are also flip-flops. They too have two stable positions: with the wings completely raised or completely lowered, and between these there is a critical, unstable point. If the wing passes this point in its upstroke it flips abruptly all the way up. If it passes it on the downstroke, it flops just as suddenly all the way down. Since the fly's wing doesn't actually snap, but with each beat one hears at most a fine 'click', it is better to call this particular bistable joint a 'click mechanism'. The flight muscles do in fact move the wings up and down but are so highly specialised that they cannot operate without the click mechanism.

Who could have stumbled upon this? We certainly noticed nothing about it during our discussion of the kinematics! It began with an accidental observation. Some American researchers wanted to see how an insect flies when it is drunk. So, for example, they put a fly in a jar with a drop of ether, and let it fly. What happens then? The animal soon falls, slightly tipsy, to the ground and there continues for some time to make characteristic uninhibited movements of 'drunken flight'. The wing-beat frequency decreases steadily, the flight tone becomes softer, until finally the fly becomes completely anaesthetised and stops beating its wings. Every insect collector is familiar with this behaviour of Diptera in the ether killing-jar. The wings stick out in any direction from the body, sometimes one way and sometimes another.

It was purely by chance that in one of these experiments someone used the cleaning fluid carbon tetrachloride (CCl_4) instead of ether. With this the fly also flew at first in drunken flight, but then the picture changed. The flight tone became not softer, but rather more powerful, rougher and less subtle. Then there followed a few irregular beats—and then suddenly the wings stopped moving. But not just anywhere; in all of the animals investigated they were in one of two extreme positions, either all the way up or all the way down. Now if one pressed on the thorax of such a fly, compressing it carefully with a forceps in alternate directions as I described in the last chapter for the tennis-ball model, it was difficult at first, but then suddenly the wing snapped abruptly from one extreme position into the other. From this experiment it may be deduced that a click mechanism is present, which, however, only becomes apparent after CCl_4 anaesthesia. It is known that CCl_4 puts 'normal' muscles into contraction, but not the highly specialised flight muscles of the flies. So one can conclude; since CCl_4 on the one hand reveals the click mechanism, and on the other causes normal muscles to contract, that the click mechanism must depend upon a contraction of such 'normal' muscles in the thorax of the fly.

The accidental experiments with CCl_4 have surprisingly given us two strong hints. First, that the wing joint works as a click mechanism. Second, that it does this only when at least one muscle contracts which does not belong to the special indirect flight muscles.

The second result comes, so to speak, as an interesting

incidental contribution. But the first result is of more interest to us. We want to look into the workings of the click mechanism in the wing joint. First let us glance at one of those ordinary, easily operated microswitches. In many types one side of the case is covered with a transparent piece of plastic, so that it is easy to watch the action of the leaf springs. In Fig. 34 it is diagrammed and explained.

Now we must learn to know a few parts of the wing joint by name. Look once again at the two extreme upstroke and downstroke positions at the bottom of Fig. 34. This is a much simplified sketch of a cross-section of the left wing joint as seen from behind. Only those parts are drawn which constitute the actual click mechanism; the many other parts concerned in setting it in motion, displacing it and providing a firm oscillatory stroke have been left out.

The shield-shaped covering plate is already familiar to us; before we descriptively called it a 'pot-lid'. It is also called a scutum, from the Latin word for 'shield'. The side plate, also called a pleural plate, is also already familiar to us. It has an inner and an outer projection. From the inner one the so-called 'musculus latus' runs diagonally downwards to the ventral plate. Between the outer projection of the pleural plate and the scutum lie the main parts of the click mechanism. We know from the microswitch that, on physical grounds, there must be at least two parts. And in fact, flies have two also, the second sclerite (*skleros* is the Greek word for 'hard'; sclerite is the name given to small hard movable bits of chitin) and the parascutum (the expression *parascutum* means 'plate lying next to the scutum'). Between the second sclerite and the parascutum lies the click mechanism, and that is where the wing with its sturdy radial vein is attached.

Let us start with the middle drawing, in which the second sclerite and the parascutum are lying in a straight line and the wing, loosely speaking, is standing out horizontally. When the musculus latus contracts, it draws the pleural wall somewhat inward. In this way it first puts the joint apparatus under mechanical tension. As long as sclerite and parascutum continue to lie exactly in line, nothing happens. But as soon as the joint between them is pressed down even a tiny bit by the movement of the indirect musculature, it happens: the system under tension springs abruptly downward as far as it can, just as in the microswitch. This flicks the wing up. Now it stands in the stable position shown at the left of Fig. 34.

Imagine that the drawing is a model, and lift the central joint up from below with your finger (the fly does it with a lever mechanism). This pushes the outer pleural wall outward and the lateral wall of the scutum inward. You have distorted and strained the system and thereby stored elastic energy. This is fine as long as the central joint does not cross the median line we have discussed (Fig. 34, middle). The instant it does, the mechanical tension begins to readjust itself, the

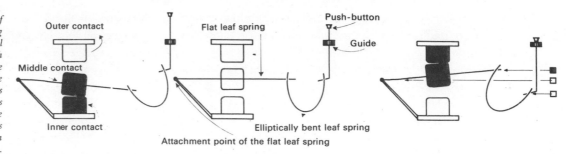

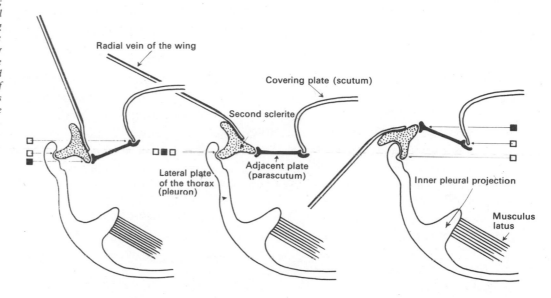

Fig. 34. Sketches showing the principle of operation of the click mechanism of the wing drive system as compared with an electrical microswitch. The poles of the switch which are in contact are black. On the left: the mechanism is switched down (first stable position). The inner joint (black square) lies below the line connecting the two outer joints (white squares). In the middle: the unstable midposition. All the joints or hinge points are in line. On the right: The mechanism has sprung upwards (second stable position). The inner joint lies above the line connecting the outer ones. The biotechnical correspondence is complete: flat leaf spring—second sclerite; elliptical leaf spring—parascutum; attachment point of the flat spring—pleural joint; attachment point of the elliptical spring—scutum joint; releasing lever—scutum. The mechanism can work only if it is under tension, that is, when the outer joints are pulled together. This is taken care of once and for all in the microswitch by the geometry of its rigid construction; in the fly it occurs as the occasion demands by contraction of the musculus latus.

parts of the joint lying between the body-wall sections move upwards rapidly as far as they can, and the wing is snatched downwards. Now it stands in the stable position shown on the right of Fig. 34. If we now press from above with a finger we bring the joint closer to the median line again, driving the walls apart, storing energy once more and thus preparing for a new upward stroke of the wing. A little more pressure, and pop—the wing is already raised again.

The parascutum and second sclerite together, then, form the 'snapping lid' of the shoe-polish tin, scutum and pleural wall form its rigid side walls. If one glues a paper wing on to one half of the lid, one can imitate the whole thing.

Furthermore, you now see why I expressly called the drawing in Fig. 31 a schematic representation. The situation is not quite as simple as it was drawn there. The indirect muscles move the lid (scutum) up and down relative to the pan (pleura plus ventral plate); however, the lid does not move the wings directly, but rather by way of a flip-flop system composed of two parts. It applies force to this system and, with a lever which is not shown, gives it a push at the right moment so that it flips over, once in the middle of the upsroke and once in the middle of the downstroke. In this way the wing oscillates up and down.

Of course the whole affair can only function if the snapping disc is under mechanical tension. If you cut away the rigid rim of the lid, this tension is no longer developed when the top is pushed. Now the lid is free to expand, there is nothing left to prevent it. For this reason no elastic restoring force can build up, and the lid no longer snaps.

Correspondingly, the click mechanism in the wing joint functions only when the surrounding muscles are stiff enough. In the fly which had been killed by ether this was not the case. All the muscles were relaxed. But in the CCl_4 flies the mechanism functioned, and we ascribed this without any doubt to the fact that a 'normal' muscle must have contracted, which was not part of the mighty indirect musculature. Now you see, too, which muscle that must be, one which is tiny in proportion to the mass of the flight muscles: the 'musculus latus'!

That is one of the many small position-setting muscles in the fly thorax. None of them have anything to do with the wing beat itself; they are not even capable of contracting 200 times a second, but are quite slow very powerful little fellows. The latus then, before the beginning of a flight, pulls parts of the pleural plates inward and holds them there. In this way it stiffens the side wall of the thorax, making it a proper bearing for the wing. When subsequently the large flight muscles begin to work, they find the necessary conditions present for them to be able to store elastic energy. Then, and only then, can the click mechanism work and set the wing in motion. A strange situation—the mighty flight machine is condemned to work in vain if it should not suit the humble little latus muscle to set the joint properly.

This always seems to me like the clutch in a high-performance racing car. You can step on the accelerator when the motor is disengaged and it will go powerfully into action, but the car does not move. It is only when one engages the clutch by a mechanical displacement which seems insignificant in comparison to the whole mighty drive system, that the racing motor is able to transmit its horsepower to the wheels, and the car shoots away. In a sense, this is the way the fly starts. As long as the musculus latus is contracted, energy of distortion is stored at each half wing beat and then suddenly released to the wing. As a consequence the wing oscillation receives a very quick violent impulse at a certain point each time. It is just as though one were to help out a heavily loaded swing by giving it a strong kick in the right direction in the middle of each half oscillation. Of course, the swing will

not whiz off like a bullet at each kick; it will increase its velocity, but slowly and smoothly. This is because of its large inertial mass which provides a compensation or damping, as in our example with the flywheel. The insect wing has no large inertial mass, but it must move a great deal of air back and forth in a short time. As a result its movement is also very much smoothed out and damped, like a spoon in a jar of honey. This is why the wing cannot follow the lightning-like release of energy so quickly and why its oscillation is so smooth, even though it is driven by a bistable flip-flop system.

If the large surface of the wing is cut away, the snapping joint can flick the small remaining stump from one stable position to the other in less than a thousandth of a second. With the wing it takes at least three times as long. so strong is the effect of the 'air brake'. Just look at the kinematic figures and follow how smoothly the downstroke is accelerated and then slowed again. The whole process occurs with such beautiful continuity that on the basis of kinematics alone we never would have come up with the idea of looking for a 'click mechanism' in the wing joint.

But if the air damping compensates for the jumping back and forth of the joint system, the flies must not actually need this highly specialised mechanism. What purpose does it serve?

We shall discover this when we look at the muscles more closely. They too are highly—one might almost say, too-highly—specialised. They would no longer be able to work together with 'normal' thoracic and joint components. Rather, they need a partner which is their peer, equally highly specialised and adapted to their exceedingly peculiar mode of operation. An exclusive club.

20. The mighty flight motor, its performance, and the problem of supplying it with oxygen

The whole thorax of the fly is crammed with the muscles of the flight motor. The remaining organs have shrunk to 15 per cent of the space at a generous estimate, leaving at least 85 per cent of the total available space to be taken up by the flight muscles. To us they appear tiny, but in relation to the

body of the fly they are indeed massively developed.

Have you ever watched a drag race? These are automobile races in which the main thing is to accelerate very rapidly at the start. To achieve this giant motors are forced into small chassis. The fly seems to me to be similarly constructed. It is obvious that the organism will not find it easy to provide the mighty flight motor with fuel and a cooling system. One is to some extent justified in calling the flight musculature a combustion motor; and as such it needs fuel and air just like any automobile or aircraft engine.

What does a motor do with its fuel? It burns it, combining it with oxygen, and in that way obtains energy. The vehicle carries its fuel along with it, while the oxygen ordinarily comes from the air, so obtaining a reliable supply of air is an important problem in high-performance motors running at high speed. In technology, turbines and compressors often become necessary when large quantities of air have to be provided in a short time. Only a rocket can do without oxygen from the air, since it carries the oxygen it needs for burning its fuel in a liquid or chemically bound form, and so can work outside the atmosphere. Nothing like this occurs in nature, but there is a whole array of clever arrangements for employing oxygen from the air.

Think of the highly specialised lungs of birds which are open at the back end, allowing the air to flow through. Such a thing is to be found nowhere else in the animal kingdom. The air is pumped through from the front to the back, then streams into large air sacs which lie beneath the skin and in many of the hollow bones, and is immediately pumped back again, crossing the lungs a second time—this time from back to front. This happens once in each breath, often in the rhythm of the wing beat. The air is 'doubly' utilised; that is, a particularly large quantity of oxygen can be extracted from it. Our lungs on the other hand are closed at the bottom, forming sacs. They do not work as efficiently as the bird lungs, but neither is it necessary that they should. Humans do not need such a high relative performance as does the flying bird—or, even more, the flying insect.

What is relative performance? This is the way one designates the energy which is converted in a certain length of time in relation to the weight of the driving system. Energy may be measured in units of heat, or calories, and so many calories converted per hour describes the power of a biological motor. In technology the power of motors is measured in horsepower, or hp, but this unit has fundamentally the same dimension as the calorie per hour, and it is purely a matter of habit whether one prefers to measure power in calories per hour or horsepower. The two can be easily converted one to the other for comparison. It is only important that the power is always given in terms of energy per unit time; that is, the energy which the motor can generate in a given time span (second, minute or hour).

Just as technologists express engine performance in relation to weight, so we in biology use the expression 'relative performance'. Again, this is just the same thing. Engine performance is usually measured in terms of horsepower per pound weight of the motor, hp/lb. A motor with a performance of 1 hp/lb, therefore, can produce one hp for every pound of its weight. If it weighs one hundred pounds it produces altogether $100 \times 1 = 100$ hp. Another, equally heavy, motor with a performance of 2 hp/lb would therefore produce $100 \times 2 = 200$ hp. In terms of energetics, which of the two would be better suited to be a flight motor? Clearly it is the second, with its higher power-to-weight-ratio, since its 100 lb generates a performance twice that of the first.

Since the builders of flying machines always have the problem of making them as powerful and simultaneously as light as possible, they are constantly chasing after high power-to-weight ratios like the devil after a poor soul. There are many tricks of construction which make it possible to extort from one pound of motor weight, as many horsepower as possible.

Power, then, is energy expenditure per unit time, measured for example in calories per hour or hp. Engine performance and the relative performance of animals are designations for one and the same thing: power per unit weight. Let us agree that in comparisons between nature and technology we shall use a metric version of the technical measurement system, hp per kg (1 kg = the mass which has a weight of 1 kg force at sea level (2.2 lb). The arrangement of Table 2 shows that the performance values of flying machines are approximately comparable, in modern technology as well as in nature, and rather high compared with those for the human. Honeybee muscles and aircraft piston motors have almost identical values. Highly-evolved piston engines possibly surpass nature a bit. Jet planes, which need unheard-of performance for supersonic flight, fall into a different category. The human being is also introduced for comparison. He comes off very badly; even the locust muscle generates ten times as much energy per unit weight. But that doesn't matter since humans do not fly, and for walking and running they need much less energy.

More than a century ago people were already experimenting with all kinds of flight motors. At first, however, they were so heavy and had such poor performance that they could only be considered for steerable balloons. Giffard, in 1852, managed for this purpose to get all of 3 hp out of a 159 kg steam engine. The performance was 0.02 hp/kg. In 1883, Gaston and Tissandier used a Siemens electric motor for their balloon; in this case the performance already amounted to 0.14 hp/kg. Around 1900 the best petrol engines were achieving only 0.07 hp/kg, but by 1906 this had already risen to about 0.5. By then they were slowly approaching the performances of biological flight motors. And gradually motors were adapted for propelling aeroplanes. One may very well say that human 'heavier-than-air' flights only became possible with the appearance of sufficiently powerful

A spiracle of the great water beetle, Dytiscus. The entrance to the tracheal system is guarded by a comb of fringed hairs which acts as a filter against dust like the oil-filter in a motor car.

and simultaneously light-weight motors. It is astonishing how technology was able to increase the performance of the piston engine by more than thirty-fold in the time between 1900 and 1960.

A century of technological history led from the first tentative attempts to fully developed constructions. The flight constructions of nature are now also, in a certain sense, fully developed. Of course, at least 350 million years have passed since the insects made their first experiments with flight!

Now let us look more closely at the way nature transports the oxygen to the motor.

If one dissects the thorax of a fly from the side and is very careful about it, one can finally remove all the dorsoventral muscles of that half of the body, until at last the powerful bundle of the longitudinal muscles has been exposed. Plate 12 shows what can be seen under the binocular microscope. The six muscle bundles, lying close together, run lengthwise through the thorax. Between them there are shiny silvery tubes which give off, at regular intervals, lateral branches which penetrate into the muscle packet. This arrangement looks like the intake manifold of an engine or compressor, which conducts the air through its branches to each cylinder. And basically, that is just what it is. These are the tubes of the tracheae, which conduct the air directly to each single little bundle of the muscle (in the first chapter we discussed these extensively). To be sure, the pipes are exceedingly thin, but they are fitted with rings and spirals of chitin like a vacuum-cleaner hose. This means that they cannot be permanently squeezed together. The shiny silver colour can be attributed to the enclosed air, which reflects the light. The branchings that we see here are still far from being the finest ones. The main trunks fork again and again, the channels become finer and finer, until finally they end in the tissue with microscopically fine terminal tubules. Often one such 'tracheole' supplies only a single muscle cell. Even the adjacent cell then has its own air hose. In insects, then, the air they breathe comes directly to the cells, and not by way of the blood as in humans.

Oxygen can diffuse into the cells from the finest ends of the tracheae, while carbon dioxide diffuses the other way round, out of the cells into the tracheae. The actual process of diffusion is rather complicated. It is enough for us now to know that the exchange of gases takes place between the cell and the tracheal endings. Thus the flight motor draws its oxygen out of the tracheal pipes and discards its waste gases into the same pipes. So the tracheal system works both as compressor and exhaust!

Naturally it must be adequately supplied with air. This happens in the thorax largely by the contraction of the flight muscles, but much more in the abdomen by a special breathing musculature which can compress the tracheae rhythmically

Table 2 Performance and types of fuel. Values have been collected from various authors.

	Performance (hp/kg)	Fuel	Comments
Human, leg muscle	0.08–0.1	Fat + sugar	Maximum possible short-term energy expenditure
Human, heart muscle	0.1–0.16	Sugar	Maximal energy expenditure
Hummingbird (Calypte), flight muscle	1.12–1.6	Sugar	Only for very brief flights
Desert locust (Schistocerca), flight musculature	0.64–1.28	Fat	Normal energy expenditure at average 'cruising speed'
Fruit fly (Drosophila), flight musculature	1.04	Fat	
Bluebottle fly (Lucilia), flight musculature	2.72	Sugar	
Honeybee (Apis), flight musculature	3.84	Sugar	
Older-type aeroplane piston motor	3.2	Aircraft fuel	
Aeroplane piston motor of modern type with variable-pitch propeller	6.4	Aircraft fuel	Short-term peak performance at maximum speed

Plate 13. *The hoverfly* Syrphus seleniticus *can often be seen hovering in sunbeams, as they strike between the leaves of the trees. This fly is able to remain poised 'on the spot', but it can also move like lightning. Its legs are held in the typical flight position: fore and middle legs forwards, hind legs stretched out backwards. Because the wings are beating at about 300 Hz the flash from an ordinary commercial flashgun (duration about 1/1000 s) gives only a blurred impression of the wings.*

Plate 14. *(Over page, left.) A cross-section through the thoracic musculature of a damselfly of the genus* Enallagma. *The big hexagonal structure in the centre of the picture is a single muscle cell, and six other cells (i.e. muscle 'fibres') surrounding it are partially shown. The radiating clear streaks are the contractile myofibrillae, and the dark enclosures between them are mitochondria. The small black spots that can clearly be seen within the fibrillae are cross-sections of the thick filaments which are shown more greatly enlarged in Plate 15. Magnification about × 9000.*

Plate 15. *(Over page, right.) Muscle fibres from the flight muscle of an insect, enlarged 400,000 times (cf Plate 14). The well-marked hexagonal arrangement of thick and thin filaments is explained in the text. This is an example of the precise geometrical arrangement that is to be found in many biological structures if they can be enlarged enough to approximate to molecular dimensions.*

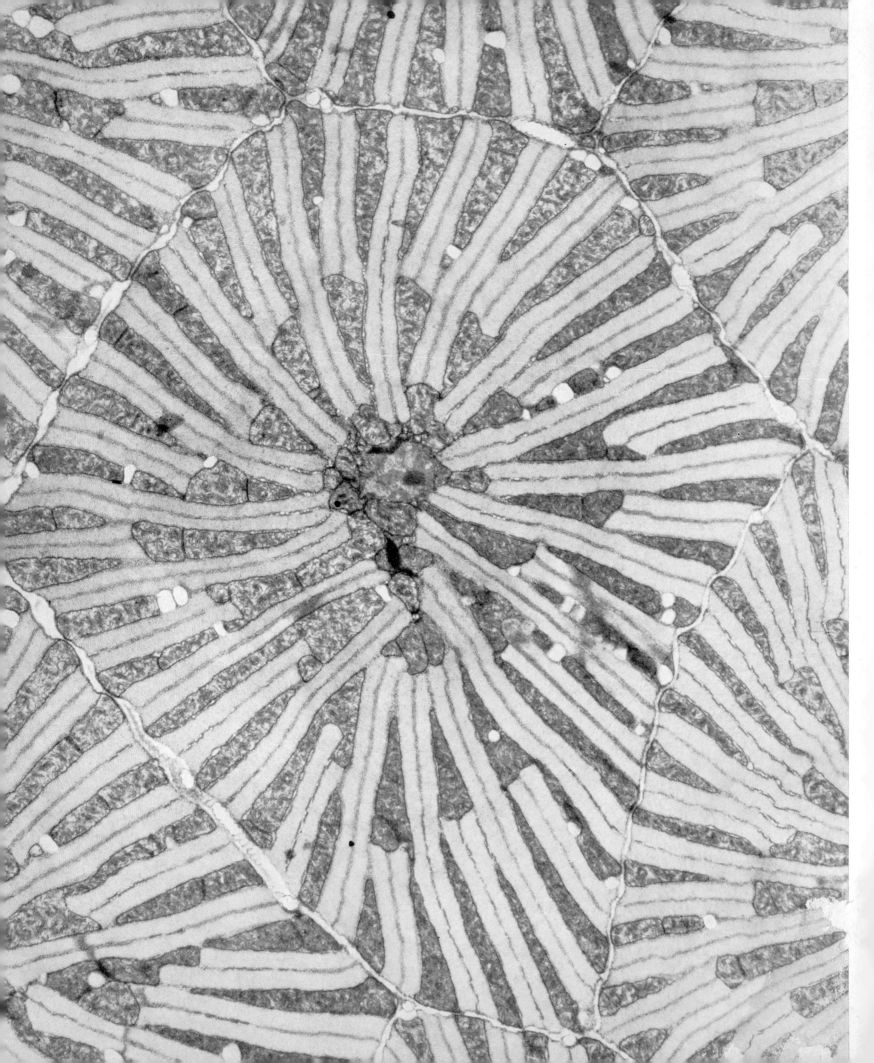

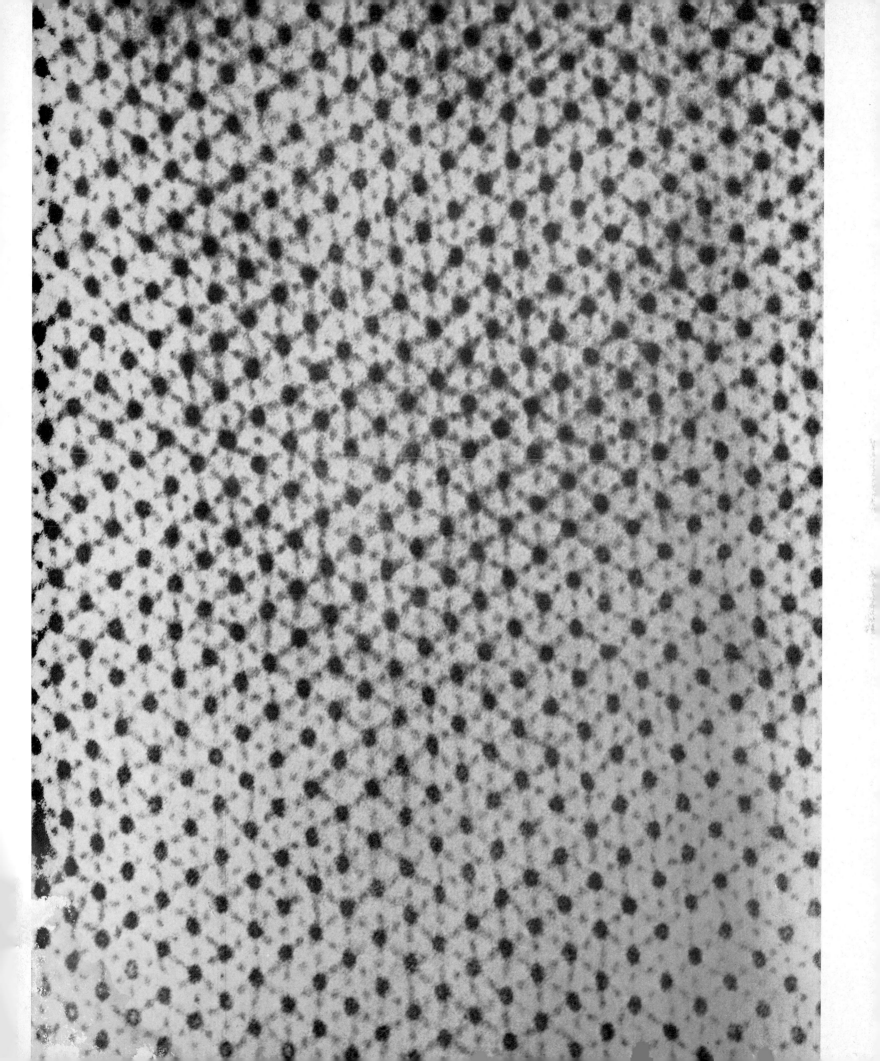

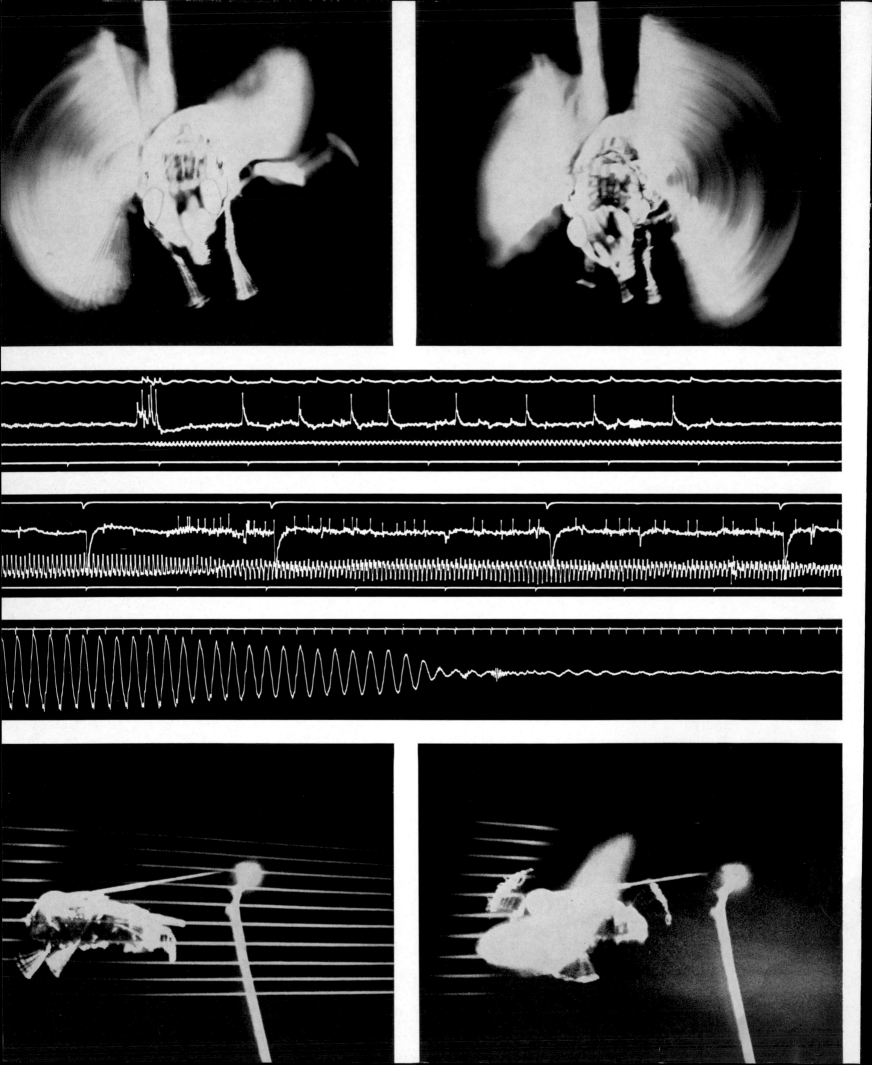

at short intervals. They expand again by their own elasticity, so they act as a pump. Sometimes, particularly in the thorax of the fly, the tracheae swell forming hollow spaces, and the rhythmic pumping of these moves especially large amounts of air. Almost all the insects pump in the rhythm of the wing beat, never slower. The dragonflies actually pump faster. They have developed a mechanism for doubling the pumping rate.

As we already know, the openings of the tracheae to the outside air are called spiracles. Dense combs of hair may be spread out in front of them to protect them from dust (see

Plate 16. Top left: *the fly Calliphora beating its right wing nearly to its full amplitude, but the left one much less so, producing a sharp turn to the left.*

Top right: *now the fly beats to full amplitude with its left wing, and to half amplitude with its right, thus producing a turn to the right (right and left apply to the fly when seen in plan view).*

First oscillogram: *record of electrophysiological reactions during a short flight. The four traces, reading from above down, show:* (a) *action potential in a left dorsoventral muscle, weakly amplified;* (b) *potential in a left longitudinal muscle (note the group of spikes at the start);* (c) *record of wing beats;* (d) *a time scale, graduated in milliseconds.*

Second oscillogram: *a recording of the changeover from 'flight straight ahead' to 'steep turn to the right'. Traces reading from above downwards, show:* (a) *action potential in a right dorsoventral muscle;* (b) *action potential in a right direct muscle, which can turn the wing back against the abdomen. (The peaks are very much more frequent here, but also much lower, so that a high degree of amplification is called for. This will be noticed in the considerable 'artefacts' induced by the potentials shown in the upper trace (upwards directed peaks), and in the marked fluttering of the baseline.);* (c) *record of the thoracic movements which take place in phase with the wing movements;* (d) *a time-scale, this time in intervals of 100 milliseconds.*
If a direct wing muscle begins to contract it turns the right wing towards the rear. Now the left wing alone continues to beat (but it will be seen that shortly after the first spike potential has appeared in the direct muscle, the thoracic movement passes through a momentary pause, into a differently phased movement). The photograph at top right shows precisely this moment of the stroke. A second later the amplitude of the right wing has fallen back completely to zero; if the fly were not tethered, but were in free flight, it would then have turned to the right.

Third oscillogram *shows a record of the wing beats during the process of cessation of flight.* (a) *a time-scale in intervals of 10 milliseconds. The wing takes about a dozen strokes to fall back from full amplitude to zero. The short, high-frequency beating at the end marks the 'clicking' of the wings back into their resting-positions by action of the retractor muscles, the same which also operate during turning flight.*

Bottom left: *a fly at rest in front of the wind tunnel.*

Bottom right: *a fly in flight just in front of the wind tunnel.*

The smoke trails in these two photographs show the movement of particular air masses, and in the last picture these are mixed together and driven downwards and backwards.

p. 74–5). Many spiracles possess closing muscles and can open or shut off their tracheae as required. In the thorax there are always two large spiracles, but in the abdomen the arrangement varies among the different families of Diptera. The thoracic spiracles are the intake openings of the flight motor; their dust-repelling hair comb has the same function as the air filter of a car, which is inserted in front of the carburettor. So a clever pumping and distribution system takes care of the vital business of providing oxygen to the machine. Now, how much oxygen does such an insect machine consume? A fly of 30 mg body weight, sitting quietly, requires about five-hundredths of a cubic centimetre of oxygen per hour. Does that number signify anything to you? If it flies, it needs about seven-tenths of a cubic centimetre in the same time. If you compare these, you will see that the energy conversion rate during constant flight at the normal velocity of two metres per second has risen to fourteen times the resting rate for just 'standing around'. Since one cubic centimetre of oxygen can burn up a certain amount, and always the same amount of a specific fuel, this means that the fuel consumption at the cruising speed of this flying machine is fourteen times higher than when it is idling. This sounds more convincing than do the individual numbers themselves.

You also will know about this phenomenon of increased energy conversion from long-distance running. When running one breathes much more deeply and rapidly than when walking slowly, to say nothing of lolling in an armchair. A greater output of energy always demands a higher consumption of fuel. While humans are capable of at most a fivefold increase in energy utilisation, the fly can raise its basic metabolic rate by a factor of 14. But that is still far from being the highest value among the insects. The bee attains the factor 19, the desert locust 25, the gadfly 26, the white cabbage moth and the cockroach 100, the may-beetle as much as 107! Even the value 500 has appeared once in the literature; but that is not really believable.

Now it is clear why the tracheal trunks are so sturdy! They must be able to bring in air a hundred times faster during flight than during rest! The highest flow rates are found in the butterflies. Let us assume that one of these rapidly fluttering insects weighs about a gram. In flight it consumes up to a hundred cubic centimetres of oxygen per hour. Now, since the air is four-fifths nitrogen, it must pump through half a litre of air per hour. On a sustained flight of ten hours, therefore, it would have transported five litres of air through its system of hair-thin tracheae, more than would fit in a gallon jug. From the point of view of the small insect, there must be a veritable hurricane roaring through its air tubes.

And this can go on for a very long time. Once someone let the fly *Phormia regina*, a relative of the bluebottle, fly in a 'roundabout': the fly was glued onto a very long needle bent into a right angle and the needle was stuck into a small vertical glass tube. The fly could sail around in a circle with the tube as its centre. Every five to six hours it was tanked up with sugar solution and then sent immediately back on its trip. These flight conditions were unnatural but useful since the revolutions could easily be counted electronically, and this made it possible to calculate how much distance had been covered. What was the result?

The fly flew for six days altogether and three minutes 'non-stop' and covered a distance of 329.75 kilometres. The mean cruising velocity was 5.5 kilometres per hour. At the end of this time the wings were torn and tattered, and the fly could no longer produce any force against the air with them. That was why it had stopped its marathon flight, and not because the fuel supply was inadequate. The tracheal system and the fuel factory, of which you are about to read, had survived this extraordinary overload, but not the wings. No wonder, they had had to swing up and down altogether 88,120,000 times! To give you an idea of this number, if you were to throw 88 million small stones into the water, one each second, day and night, it would take you about three years!

21. Fuels for insect flight

The system of pipes, then, can supply the respired air in quantities sufficient even for cases of extreme need. It is constructed with a margin of safety. The insect body never lacks oxygen, even during the fastest and longest flights. But where does the fuel come from? In fact, what is the fuel for insect flight?

The first question is soon answered: from food, of course. Nutrient-containing substances are taken up by the mouth-parts and digested in the middle section of the digestive tract; that is, they are split up into easily soluble products. These are immediately resorbed in the midgut and transported by the body-fluid system of the insect to other parts of the body. The food can contain proteins, fats and carbohydrates (for example, sugar and plant starches). It is the products of the breakdown of these three groups of substances which are taken up by the midgut. From these products new substances can be synthesised again and reserved in special storage tissues for later use. These end products belong again in principle to the same categories—fats, proteins and carbohydrates—though generally as different chemical compounds. Fats are very often synthesised; the place where they are stored is called fatty tissue or the fat body in the abdomen.

And this brings us to the second question: to the type of fuel. In many orders of insects it is just this fat which serves as fuel for the flight motor: the fat body is equivalent to a fuel tank. The migratory locusts belong to this group of 'fat burners', as do the butterflies and most of the beetles.

Protein is practically never used as a fuel; only the most catastrophic lack of nutrients could cause this source to be touched. Carbohydrates, on the other hand, are the most commonly used fuels for the flight motors of flies and bees, as well as for the muscles of humans and other mammals. Compared with fat as a flight fuel they have both advantages and disadvantages. The advantages come from the fact that they are water soluble and therefore present in the body fluid which bathes the muscles. They are therefore immediately available. Fat, in contrast, must first be broken down and delivered from the distant fuel tank. This takes some time. But water solubility is also the greatest disadvantage of the carbohydrates: only as much sugar can be dissolved as the small volume of body fluid allows. This means that for the 'sugar burners' the supply of fuel is very limited; therefore they must invent a sort of chemical factory to replenish the store continually. This is the more necessary since they consume more than a dozen times as much fuel in flight than they do when at rest. Moreover, the supply of energy contained in a milligram of sugar is much less than that in a milligram of fat: sugar is 'regular' grade petrol, and fat, 'super' grade. H. Steiner summarised all this so well in a newspaper article that I shall quote him verbatim:

'Energy production in the biochemical cycle of the cell can be achieved by way of either fat or sugar. So far, however, it has been supposed that insects, like mammals, burn sugar as a fuel during flight. The example of the honey-bee appeared to justify this assumption: it supports its flight with the aid of its blood sugar and above all with the supply of nectar in its honey stomach, which permits it a flight duration of 15 minutes and a radius of action of seven kilometres. The fruit fly *(Drosophila)* flies with sugar as a fuel: if one is made to fly for about five hours its tank becomes empty and it must land and is completely exhausted. A tiny drop of syrup can revive it immediately and allow it to fly on for some time. Nevertheless, sugar is obviously not a very practical flight fuel: it can only be utilised in a water solution, which means that the animal must carry along an additional "water ballast" for each gram of fuel. Moreover, the energy content of sugar in comparison to the same amount by weight of fat is considerably smaller. An insect flight machine which requires blood sugar (glycogen) must transport roughly eight times as much weight as if it used fat, to achieve the same performance.

Thus the insects which are adapted to sustain their flight with fat (as the most recent investigations have shown) are those which carry out long migrations, for which every gram of fuel that has to be carried is important. Naturally the migratory locusts and butterflies, some of which are known to cover very long distances, belong to this group. The use of fat as a fuel is so efficient here that a migratory locust in one hour of flight consumes only 0.8 per cent of its body weight, while a bee requires 10–30 per cent of its body weight for the same time. It must fill its tank correspondingly often and cannot go far away from its flight base.'

So far we have been so much occupied with our domestic animal, the fly, that we ought to finish by looking a bit more closely at the way it manages its fuel supply. This time let us also consider a specific kind of fly, a small, shiny golden-green animal with the pretty name *Phormia regina*. *Phormia* burns sugar as fuel, a particular sort of sugar called trehalose (there are also many other kinds of sugars).

The maximum amount of trehalose that will dissolve in the body fluid is 30 grams per litre, so it can be quite concentrated. If one lets *Phormia* fly for three hours, it slowly tires. One can tell this from the gradual decrease of the wing-beat frequency. The trehalose level also falls off continuously from 30 grams per litre shortly before the flight to less than 7 grams per litre towards the end of the flight. Thus the sugar concentration is reduced to at least one quarter of the original value. From this it may be concluded that the level of tre-halose at any time determines the wing-beat frequency and capacity for flight. That this conclusion is correct is shown by control experiments. For if one injects a half-exhausted fly

with half a cubic millimetre of this solution (a problem in itself, you may be sure!), the wing-beat frequency and 'willingness to fly' immediately shoot up again. The higher the trehalose concentration in the half cubic millimetre of water injected, the higher the frequency goes. If the fly is allowed to drink its fill at last, it flies again with maximal frequency, as it did at the beginning. But the higher the frequency, the greater the expenditure of energy. This demonstrates that the trehalose concentration determines directly the energy expenditure of the muscle at any time. And this seems quite reasonable; the less energy-providing substance there is, the more parsimoniously it is spent. But that cannot go on indefinitely; the motor must eventually stop. And in any case the reserves soon become inadequate for such energy-consuming manoeuvres as climbing, for which a high wing-beat frequency is required. The supply of trehalose must continually be renewed so that its level never sinks too low, accordingly, somewhere in the body of the fly there must be a trehalose factory. And in fact there is one; it is, of all things, the fat body!

You see, the body fluid contains in solution not only trehalose, but also glucose, which the flight muscles of *Phormia* are apparently not as willing to utilise as trehalose. The glucose is taken up by the fat body, which synthesises from it the proper fuel, trehalose, and returns this to the blood. In technology this process is called 'refining' the fuel. A motor specially developed for flight does not run on ordinary automobile petrol, but requires special aircraft fuel. The oil refinery turns raw gasoline into this sort of light aircraft fuel. Something similar happens in the factory of the fat body. Together with the raw material, the flies carry along with them, as it were, their own refinery, which continually synthesises fuel sugar to replenish the supply when it becomes dangerously low. Each milligram of the fat body of *Phormia* can synthesise 300 micrograms of trehalose in an hour. Approximately 200 μg can be stored at once in the total blood volume of the fly. In normal flight the fly consumes about 15 μg/min. Without stores of glucose and without the ingenious sugar factory of the fat body it could fly for only 13 minutes (200/15), then it would have to stop to fill its tank.

The honeybee is in a very similar situation, as we know. Its fuel depot is primarily the nectar in the honey stomach. This tank is empty after 15 minutes of flight. If the bee then does not want to mobilise other reserve substances, it must stop and fill up. The large migratory locusts, on the other hand, as fat burners, can fly for many hours—without a built-in refinery! Their reserve substance is so loaded with energy, and at the same time so light, that 10 mg are sufficient for an hour's flight.

If the oxygen-supply system of the tracheae was seen to be excellently constructed, then the fuel supply seems even more ingeniously controlled. In the muscle, oxygen and fuel meet; combustion delivers the energy necessary for flight. Similarly a bit of wood which has been set alight delivers energy which one can for example use to boil tea. In both cases it is a matter of combustion, and from the viewpoint of energetics the processes are comparable throughout. Chemical combustion—the details of the reactions are fantastically complicated—takes place in specific parts of the muscle. The muscle burns fuel with the aid of oxygen. Waste gases such as carbon dioxide and other substances are left over at the end. Precisely how it all happens—God and the biochemists may know.

Chemical combustion is one of the most important basic processes in life's great workshop. It delivers energy, and without energy there could be no flight, no movement, no life.

22. Cooling the engine

In the Sacramento Valley in California they have a saying that an automobile engine is only as good as its cooling system. With a summer temperature of 120 °F (49 °C) this may well be true. At one time the cooling system of the engine was one of its most troublesome features, but today the radiator is made so large that it is possible to drive through Death Valley without straining the engine—and that is one of the hottest places on earth. The technologists have solved the problem by transferring the heat to a fluid, and this fluid is then cooled by a stream of air. In other words they have placed an intermediate step between the engine and the air. Air-cooled engines do not have this intermediate step. What about insects? So far we have found a biological analogue for every invention of technology, but in this case there is none. We might then ask if it is unnecessary for the flight motor of insects to be cooled.

Let us take as an example a large nocturnal butterfly of the family *Celerio*. It is placed in a cool room in which the ambient temperature is perhaps 16 °C, and prodded until it is sufficiently annoyed to fly away. However, we have previously placed in its thorax a minute electrical thermistor, so small, as fine as the finest needle, that it does not disturb the butterfly at all. The temperature can be measured directly in degrees centigrade. At first the butterfly has the same temperature as its surroundings, 16°. Soon it begins to vibrate its wings, at first almost invisibly, and then with ever increasing strength. It is warming itself up; the muscles are turning chemical energy into heat. In a few minutes the temperature of the thorax rises from 22° to 38°, and then the butterfly seeks a suitable launching place and flies off.

During flight the muscle temperature remains steady at 38°, but shortly after take-off it may rise as high as 40° for a short while. As soon as the butterfly lands, the temperature starts to return slowly to that of the surroundings. But as long as the butterfly is in the mood for flying, it starts to vibrate its wings whenever the temperature falls to 34.8° to keep the temperature above this value. The temperature regulation works in the following way: the heat production can be switched on and off according to the work being done by the muscles, the switching temperature being 34.8°. In this way the temperature can be held at about 38°, the temperature during steady flight, although there is no cooling system. The details of this mechanism are not known at the moment.

Because these animals are nocturnal, they lose a considerable amount of heat by radiation, and also by conduction to the cool air. It is possible that there is an equilibrium between the heat losses by these methods and the heat production, such that the temperature of the insect is about 38 °C. In any case the musculature cannot tolerate a temperature much higher than this, because its structural proteins would be damaged.

Migrating locusts cannot use this simple method. The surrounding air is warm, their bodies are heated by the rays of the sun, and in addition the flight muscles generate a considerable amount of heat. They lose some heat by evaporation, by allowing the surface of their bodies to become damp, then the heat required to evaporate the moisture is taken from the surface of the insect, which is thereby cooled. This is an old trick used by Mediterranean peoples who make their water jars out of a porous clay. There is always a little water seeping through to the outside of the jar, and this keeps it beautifully cool. In the locust, this evaporation may reduce the temperature difference between the flight motor and the surroundings by 10–20 per cent, and is not very effective. A further 10–20 per cent is due to radiation, but most of the heat, 60–80 per cent, is lost by direct thermal contact between the surface of the thorax and the air. This is the most important way in which the insect loses heat, and it works particularly well if the surface of the thorax is smooth. Locusts, dragon-flies, and many flies have a mirror-like thorax. A layer of hair, on the other hand, is designed to insulate against heat losses like a fur coat. Insects which do not have to worry about getting rid of excess heat, but which are concerned with heat conservation, have a coat of hair on the thorax. Many twilight and night flying insects fall into this class.

So the circulation of the blood and ventilation through the tracheoles contribute very little to the cooling—most of the heat is simply carried away by the layer of air which makes direct contact with the surface of the thorax. Nonetheless the cooling is so efficient that only about 5–15 per cent of the excess heat flows to the head, abdomen and legs. In these parts of the body the temperature of the surrounding air is in control. In nocturnal butterflies this may be 15 °C, whilst a couple of millimetres away the temperature of the thorax is 38 °C.

Insects always keep a cool head.

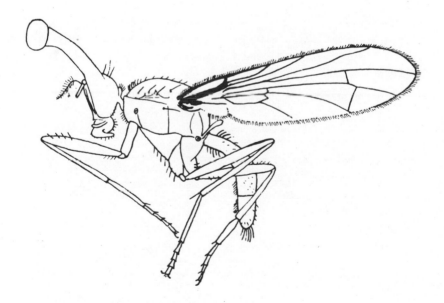

23. The wonderful substance resilin

In order to understand the very high efficiency of insect flight, we must examine the elastic component of the vibrating system. What does elastic mean?

Perhaps you have heard of a new type of ball made in America out of a highly compressed black-rubber. These balls are nearly perfect elastic bodies. If they are dropped from a table top they would bounce back almost to the height of the table, but not quite. After the ball has fallen and is about to make contact with the floor, it has a certain amount of kinetic energy. When it hits the ground it is deformed elastically and all the kinetic energy is transformed into strain energy, except for a small, but unavoidable, amount which is lost as heat. When the ball comes to a complete stop, the process is reversed: the elastic material begins to return to its original shape and in the process pushes the ball upwards again. A small amount of heat is lost once more, but most of the stored energy—that is the important word, stored—is transformed back into kinetic energy. The ball springs back up almost to the point from which it started, but, as we have said, not quite to the starting point since the ball is not completely elastic and some energy is lost as heat, and air resistance also slows it down.

The process is repeated. At each bounce some of the energy is lost, and when finally it is all dissipated, the ball remains stationary. The more perfectly elastic the ball is, the less energy it loses with each impact, and the longer it goes on bouncing. This toy leads us to a new definition of elasticity. A body is more elastic, if, in the process of transforming energy from one form to another, it has lower losses. In other words, a highly elastic body can store kinetic energy by deformation with little loss, and then give the energy out again as kinetic energy, again with only a small loss.

There is no such thing as a perfectly elastic body, as a certain amount of loss is inevitable with energy transfer. If there were such a thing, we could define it in the following way: a perfectly elastic body gives back exactly the same amount of energy which has been put into it.

We have talked so far about a rubber ball which gains its kinetic energy by falling. We could also pivot a hammer at the end of its handle, so that it could tip either way and strike the floor, like the clapper in a bell. Now let's fix the ball to the floor at the point at which the hammer would strike. It works like a spring, and when the hammer hits it it bounces back up again. The hammer goes on bouncing up and down with ever decreasing amplitude. We could have used a spring instead of the ball, but it would not be so elastic, and so would not allow the hammer to bounce so many times before coming to rest. Imagine that the hammer is connected to a reversible motor, something like Fig. 35. The hammer is beating on a concrete floor, first one side and then the other, the motor reversing its direction every time the hammer strikes. This is possible, but it needs a lot of power, perhaps 500 watts. Try it with your wrist, and you will soon find out how much energy it takes. After each change of direction the head of the hammer must be accelerated from rest, because the whole of the kinetic energy of the last swing has been absorbed by the inelastic floor. We have generated some heat and some noise, but not stored any energy.

Now let us attach one of the black balls to the floor where the hammer strikes on each side. The hammer will go more quickly and is much easier to move, the power required being maybe only 150 watts. The balls store the energy of the falling hammer, and give most of it back again to start it on its upwards journey. No longer does the motor have to work to accelerate the mass of the hammer, it only needs to replace the energy which is necessarily lost by the balls, absorbed by friction or is lost due to air resistance.

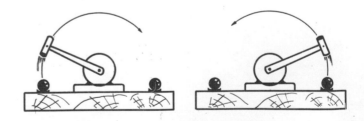

Fig. 35.   The storage of energy in an elastic material, and its effect on a periodically repeated movement. Explanation in the text.

This is the advantage of an oscillating system which has a high elasticity. Once it has been set in motion, it only needs a small amount of energy to keep it going, which compensates for the unavoidable damping. If the damping is small, that is the energy storage is working well, the power necessary is small indeed.

The insect-wing system is just like this. The highly elastic energy-storing components are the flight muscles, chitin walls of the thorax and the wing suspension. This system has been very thoroughly studied in a large migrating insect, the desert locust. It stores most of the energy of the down-stroke and releases it again on the upstroke. Similarly, it stores most of the energy of the upstroke, and delivers it to the wings again on the downstroke. Altogether about 86 per cent of the energy is stored, and only 14 per cent has to be supplied by the flight motor.

A relatively large proportion of this energy, about 25 per cent, is stored in the minute wing suspension itself. Although the system is very small, about the size of a pin-head, it is effective. It stores about 96 per cent of the kinetic energy delivered to it! The locust probably could not mi-grate without it. The strong thoracic musculature would not be powerful enough to accelerate the wings afresh at each wing beat. The high wing-beat frequency would not be possible either.

Technology has been unable to produce a substance with such good elastic properties as the wonder substance of the basal hinge—resilin. It is possible to cut out a piece of resilin and keep it stretched at three times its natural length for three months. If it is then released, it returns immediately to exactly its original length. This cannot be done with any other substance produced naturally, or with synthetic rubber. A piece of rubber will contract again when it is released, but not to its original length—it will be a little longer than before, since it is not perfectly elastic.

Morphologically, resilin is a highly specialised form of cuticle. This is the name given to the outermost layer of the bodies of arthropods, which chemically is a protein, with enormously long molecules, up to a millimetre long, and ranking as giant molecules. What is more important is that these molecules are cross-linked to each other at various points along their length, thus forming a net, like vulcanised rubber. The cross-linking is extremely regular, so that each individual layer of resilin appears from all directions to be a very orderly array of fibres and cross-links. This is the secret of the almost impossibly high elasticity. The layers are from two- to five-thousandths of a millimetre thick, and they lie on top of one another like the layers of an onion. Between each two layers of resilin is a very thin (1/5000 mm) lamella of chitin.

Since the discovery of resilin, scientists have been fran-tically at work, trying to make a similar substance with 96 per cent elastic efficiency. It would be an extremely useful material. It seems that technology does indeed have some-thing to learn from nature.

Resilin has also been found in the wing hinges of dragon-flies, it even occurs in crabs, and it is certain that similar liga-ments are to be found in most insects. The suggestion has been made that it might be present in the wing hinges of birds and bats, and it has also been known for a long time that there are energy-storing ligaments in horses' feet. No-where else, however, is such an efficient substance to be found as in the insects. The insect physiologist Torkel Weis-Fogh and his pupils have discovered and investigated resilin. Our present knowledge of the flight of the locust is largely due to their careful research.

24. The construction of the thorax, light yet strong

However lightly the flight motor is built, this is of no avail if it is enclosed in a massive thoracic box. For powered flight everything must be as light as possible, yet strong at the same time: motor, wings, fuselage, mechanical linkages and so on.

Soaring flight calls for very special requirements, with every-thing being extremely light in order to give good gliding characteristics. Lightness, economy of materials, and stability must be combined by building up a framework out of thin rods and tubes. The Deutsche Museum in Munich exhibits pieces from the old Zeppelins, which consisted of an outer skin stretched over a confusing tangle of thin pieces of aluminium. There is also on display a cross-section from a Ju52, and the same applies to that: interlacing struts and light metal tubes built up into a framework which was covered with a corrugated skin.

But to return to gliding flight. Directly below the Ju52 is displayed the structure of the cockpit of a glider, which shows how a complex of struts can be carefully built up to achieve maximum stability and resistance to distortion, with the minimum of material. This framework is then covered with doped canvas. The military aircraft of the First World War were constructed on a similar framework, and a cross-section of a Fieseler–Storch makes use of such a one.

Let us look now at the construction of the thorax of a typical insect. One of the most strongly constructed is the thorax of one of the big dragonflies. Fig. 36 shows a diagrammatic reconstruction; the lightness of construction is striking. Here there is no question of a tubular framework covered with fabric, but a system of light, but relatively strong, sclerotised ribs, overlaid with a thin membrane of the finest epidermis. It will be seen that there is a kind of 'mezzanine floor', called the *furca,* upon which the massive flight muscles are mounted. The lightly built, double ribs are not straight, but somewhat bowed outwards, and are joined together above and below by longitudinal rods, thus giving a certain amount of elasticity against longitudinal stresses. The four ribs shown in the diagram end above in short, hammer-shaped longitudinal processes, and it is upon these that the bases of the two pairs of wings are mounted. All these items clearly combine to give a light form of construction. The strongly sclerotised rods take the stresses and bear the weights when the flight muscles of the wings come into action. Like the fabric of the Fieseler–Storch, the thinly sclerotised covering material is only a closure, and has no supporting function.

The wing processes, the furca and the longitudinal rods all serve as attachments for the flight muscles. Each of the four wings has its own individual direct muscles, and dragonflies can even move each wing independently of the others: this is one reason why they are such acrobatic fliers. They are able to stop in mid-air, and even to fly slowly backwards. They can make the steepest turns, and reverse direction instantaneously. Thus they are able to pursue their prey however frantically it may try to dodge them. There are some insects that can fly more quickly than dragonflies, and many others that can carry on longer in sustained flight, but the control and mobility of dragonflies is unrivalled. It is understandable that each wing, with its own musculature, needs its own separate wing process or fulcrum, but at the same time these

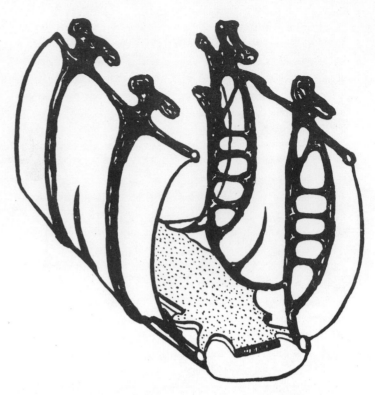

Fig. 36. *Thorax of the dragonfly* Aeschna cyanea, *seen obliquely from in front and above, cut off anteriorly and posteriorly and with wings and upper surface removed as well as the internal organs. The sclerotised cuticle is shown white, and the furca is stippled. For exaplanation, see text.*

processes must be tied together longitudinally so that they cannot slip towards each other. They must also be sufficiently rigid, and this rigidity is achieved, as in man-made machines, by linking the two processes of one side together.

A thorax constructed in this way leaves hardly anything more to be desired, with structure and function complementing each other perfectly. Locusts and flies also have similar wing processes, but in these insects the wings are firmly co-ordinated, and cannot be operated independently however much the insects might try to do so. Hence the thorax of a locust, and even more that of a fly, can be built up from stiff, though thin, plates. The distinction between structural skeleton and superficial covering membrane is not so clearly drawn in these insects. Dragonflies in contrast are specialists in light structural frameworks.

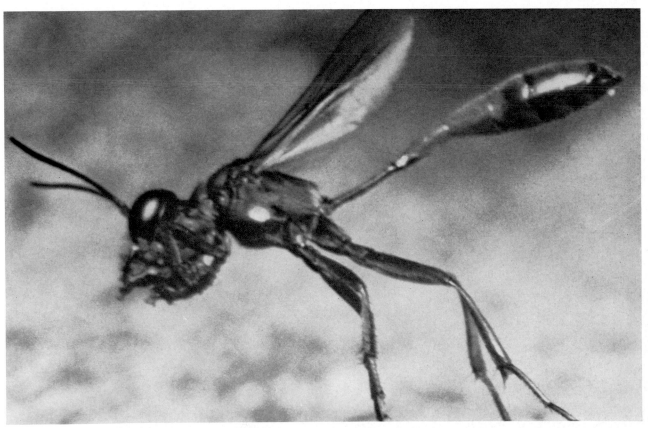

25. The deer botfly is not a living projectile, and it does not fly at supersonic speed

By now we have had so much to say about the flight motors of insects that it is time for us to see what these motors can actually achieve. A preliminary measure of the performance of a flight motor is the velocity that it can generate in the flying object. The more powerful the motor, the greater the velocity can be. There are better and more logical ways of expressing the efficiency of a motor, but to the driver of a car, and the pilot of an aeroplane, velocity is such a tried and trusted criterion that we must begin with the question: how fast can insects actually fly?

It may seem incredible, but until just about ten years ago it was not possible to find reliable data about this, even for

Table 3 brings together a few apparently reliable figures from different authors: the longer the black line, the greater the velocity. These are average speeds over a distance, and not momentary peaks lasting one second, or a few seconds at most. Sudden bursts of this sort might reach maxima that were from 20 to 100 per cent higher than the average speed, and even these are not so very remarkable. There is little serious evidence for the existence of 'lightning' insects such as we meet in various legends and science fiction. The fastest flying insects are the big dragonflies, hornets, bees, hawk moths and botflies, but even these, with their average of 20–30 km/h are only travelling with the speed of an automobile in urban driving. Bees can raise their speeds to 29 km/h after they have fuelled themselves with an especially concentrated sugar solution. The biggest dragonflies can momentarily speed up to 60 km/h and hawk moths perhaps to 50 km/h but these are probably the highest speeds attainable by any insects. So the fastest insects, under the best conditions can only fly as fast as a starling for a few seconds at a time.

The picture looks slightly different if we calculate how many lengths of its own body the insect leaves behind it, and this method helps a little to remove the inequalities that arise when we compare the speeds of little and big animals in the same units. Right away it becomes obvious that tiny little insects cannot fly as quickly as the giant dragonflies, even though they seem to have an equally well-organised flight system. Some very remarkable figures emerge at this point. If a blowfly 11.5 mm long flies at a speed of 3 m/s it leaves behind 261 body lengths every second, and this is a high

the honeybee, an insect that has been studied by so many people, beekeepers, scientists and naturalists. The reason for this lack of information is that insects rarely fly for long at a constant speed, except when they are migrating, and furthermore their speed over the ground can be falsified by fluctuations in wind speed. If a bee, with a flight speed of 5 m/s flies in a following wind of only 2.5 m/s (a very light breeze, hardly 10 km/h) the ground speed of the bee is already increased by 50 per cent up to 7.5 m/s. This last is the speed that would be calculated if the bee were timed with a stopwatch as it flew between two points, yet it would be too high by half.

figure. Compare this with the following: the fastest mammal, the cheetah, when in pursuit of game, achieves 18 lengths per second, a horse 6, and a man, in a burst of speed, 5 at most. A Volkswagen at motorway speeds achieves 5–6 lengths per second, a swan less than seven, and a starling about 80; the speedy swift in a dive 60. In contrast a sporting aeroplane reaches only 10, a small jet fighter at sonic speeds somewhat more than 30, and, at three times the speed of sound, 100. The fly clocks up an astounding 250–300 body lengths per second, outstripping the fastest bird by two to one, and the subsonic airliner tenfold. Comparisons like these show again how cautious one must be, and critical when setting up a natural process against an artificial one.

Moreover by this process of expressing speed in terms of number of body lengths covered per second the smallest insects come off best, and their 'relative speed' is the highest of all. Little wonder, therefore, that their demand for fuel is very high, and can rise to one hundred times as much as they need during a period of rest. Still higher velocities are clearly inhibited by the problem of producing a big enough turnover of energy-generating substances. Although many of these facts were already known long before the Second World War this did not prevent a false report from gaining currency all over the world. A biologist had claimed that a deer botfly could reach supersonic speeds, and an aerodynamist had proved the veracity of this report by mathematical calculation. What truth was there in it?

In an article in the *Proceedings of the New York Entomological Society* there appeared the following statement:

'On the other hand, on 12,000 ft summits in New Mexico,

Plate 18. A sand wasp of the genus Ammophila *in free flight. The abdomen is held up in the characteristic way and the attitude of the legs, outstretched and hanging down, shows that the photograph was taken either shortly after the start of the flight, or just before landing.*

Table 3 Flight speeds of insects, values taken from various authors.

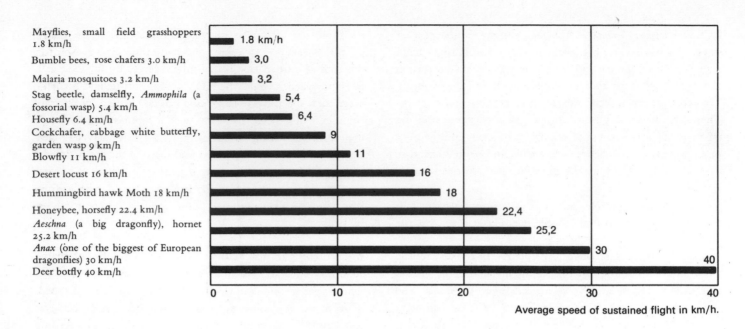

Average speed of sustained flight in km/h.

I have seen pass me at an incredible velocity what were quite certainly the males of *Cephenomyia pratti*. I could barely distinguish that something had passed—only a brownish blur in the air of about the right size for these flies, and without sense of form. As closely as I can estimate, their speed must have approximated 400 yards per second' [= 1300 km/h]

This speed was about 10 per cent above the speed of sound at sea level, and if the lowest possible value is taken for the drag that would be set up at such velocities by the fly's body alone, not counting the wings, it would amount to about 100 grams. The air pressure thus generated against the front end of the insect, i.e. its head, could not be less than half an atmosphere and this would have squashed the head in. This would be a bad look out for the poor fly! An estimate of the effort that the fly must put up gives a completely convincing rebuttal to the claim: it would require 370 watts, or half a horsepower, as much as an electric iron, or a small refrigerator, or six average light bulbs. To generate so much energy the insect must take in, *every single second*, 0.3 grams of energy-producing food, or one-and-a-half times its own weight.

Botflies occasionally strike against the human skin, and they produce a noticeable impact. It might be calculated what would happen if a botfly were to approach at supersonic speed, and came to a stop on the skin in a distance of one centimetre. In spite of the tiny mass of the fly (0.2 g) the impact would still generate a force of 140 kilograms, and this is the weight of a motor cycle! If this were true, no-one would dare to allow a botfly to land on the skin, where its impact would have the effect of a heavy calibre pistol bullet.

Irving Langmuir, who made this calculation in 1938, did not content himself with merely reducing the tale of supersonic flight to an absurdity, but he also checked the circumstances in which the observation was said to have been made. He carried out an experiment with small objects the size of a fly, and found at what velocity they assumed the appearance of 'a brownish blur in the air'. Upper and lower limits can easily be determined for this speed, and it can also be found out at what point traces remain, and when nothing at all can be seen. The figure arrived at was about 40 km/h. The fly's effort at that speed is 3.4 mW, and this amount of energy can be generated if the fly can take in about 5 per cent of its own body weight of food per hour. This agrees well with wind-tunnel measurements, and is certainly of the correct order of magnitude.

So the deer botfly was quietly struck off the list of the fastest flying machines as long ago as 1938, though it continued to be quoted in popular reference books long after that. Sad, because it would have been a world record at that time.

If body length is taken as criterion, the pigmy shrew is 120 times smaller than the African elephant—excluding the trunk! Comparing wing spans in birds, the hummingbird and the South American condor are in the ratio of about 1:50, but among insects the range of size between expert fliers is much greater than this. The giant dragonfly *Megaloprepus coerulatus,* with a wing-span of 18 cm, and the fairy fly, *Alaptus magnanimus* (Hymenoptera, Mymaridae) with about 0.25 mm are in the ratio of about 720:1! The all-out extreme is reached by comparing the 40 cm length of a single wing of a fossil dragonfly from the Devonian period in North America with the 0.075 mm of the living mymarid *Patasson crassicornis,* a ratio of 5300:1. We have already seen, in our discussion of the gliding flight of butterflies, that gliding characteristics change very considerably as body size is altered. For this reason we cannot directly compare the sail butterfly with a big man-made glider, although the size ratio between these two is only about 100:1. Indeed the same is true even between model gliders and real gliders, only ten times bigger, and of similar aerodynamic design. What a difference there must be, then, between the biggest insects and their diminutive relations. How indeed, do these pigmy insects manage to fly at all?

Before the function of an organ can be profitably considered its structure must be precisely studied, so let us compare the wings of outstandingly small insects belonging to different orders, for example from the Coleoptera, Thysanoptera and Hymenoptera. Two points emerge: firstly that these insects are at most one-tenth of a millimetre long, and secondly that they are all incredibly similar in appearance. Each wing consists of a club-shaped main stem, which carries a fringe of erect hairs lying like a fan in one plane. Naturally the beetle has only one pair of wings of this type, since the fore wings of beetles function as wing cases, or elytra; and, equally obviously, Thysanoptera and Hymenoptera will have two pairs of such wings. The individual wings of this group,

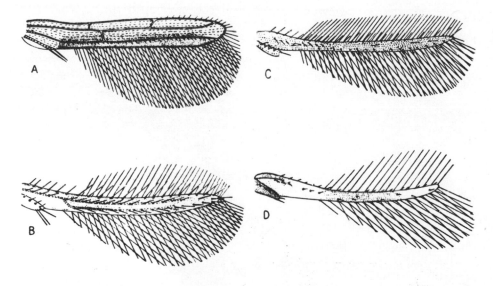

Fig. 37. Bristly wings of various species of Thysanoptera, each time the right fore wing is shown. A, Aelothrips kuwanii; *B,* Euthrips ulicis; *C,* Taeniothrips inconsequens; *D,* Euthrips orchidi. *The length of the wings is only a few tenths of a millimetre.*

however, are quite different in plan from any type of wing that we have studied so far. Moreover they show close resemblances between themselves, though with variations. The same is found to be true if we compare the wings of different diminutive species belonging to the same order of insects: for example, Fig. 37 shows the wings of four different thrips of the order Thysanoptera.

These resemblances, especially those between insects belonging to different orders, are said to be 'convergent forms'. A convergence in the animal kingdom very often indicates that forms of quite different ancestry, and independent evolution, perform similar functions. Mammals and insects really have very little relationship with each other, yet the fore limbs of moles and of mole-crickets are astonishingly alike. They share the same function, that of digging the soil, and for this activity a particular design is most suitable. Over millions of years both groups have evolved such a shovel-like limb, an example of structural convergence. Anywhere that convergent forms are found it can be assumed with comparative certainty that they function in similar fashion, and perform similar tasks. As far as miniature wings are concerned, we can infer from this that the tiniest members of all the orders of insects fly according to the same principle. But what is this principle?

In an earlier chapter, when we analysed the normal, lifting type of wing by means of polar co-ordinates, we saw that the theory leads to a significant conclusion: that the smaller the wing and the more slowly it beats, the less efficiently it works, both statically and dynamically. Thrust and lift grow less, and the undesirable drag grows bigger and bigger. This has nothing to do with better-or-worse design of the individual wings, but is simply a consequence of the physical forces which govern the airflow. Polar co-ordinate analysis of the flight of the vinegar fly, *Drosophila,* has shown that small insects can indeed fly at large angles of attack, and are therefore less liable to stall and less sensitive to sudden gusts, but they achieve this stability at the cost of a greater expenditure of energy because the unavoidable increase in drag absorbs a high proportion of the available motive power. They are like a bicycle that needs oiling, so that the rider has to toil very hard, and uses up half his effort in just getting the cycle moving.

We know from the study of aeroplane wings that a well-designed profile with suitable upper curvature gives a good lift component, allied with minimum drag. The smaller the wing is made, and the more slowly it is pulled through the air, the more rapidly the lift falls off and the drag increases. The profile of the wing has less and less effect, and indeed in small wings actually begins to be a hindrance rather than a help. For this reason both natural and artificial wings depart from this construction when it comes to small, slow-moving flying machines. The wings of a balsa-wood glider no longer have a 'profile', but are the same thickness all through—though they still have a slight upwards curvature—while small paper aeroplanes have flat wings of uniform thickness. It is just the same in the animal kingdom. Big gliders, such as the condor, lammergeier, albatross or buzzard all have wings that are thick, strongly arched, and superbly profiled. Even the domestic pigeon goes along with this, but the small warblers already have wings that are much thinner and much less arched. The wing of a hummingbird is symmetrical in profile, and therefore practically without any upward arching. The wings of butterflies are already paper-thin, and those of smaller insects are certainly so.

All the creatures mentioned so far, however, fly according to 'orthodox' principles, generating plenty of lift component, and only moderate drag, but all the time, as we go down in size, the difficulties increase. Small flies need to generate a very great deal of power, and that is why they have such massive flight muscles.

When we come to pigmy insects, the tale becomes very involved. At these tiny dimensions the laws of airflow mean that beating wings generate hardly any lifting force, but an extraordinarily high drag, and so we come to the end of the road for the classic form of wing. What do we do now? Adapt, come to terms with the facts, and devise some other system. Big insects were the first fliers in the history of the world, but then insects became progressively smaller. For big insects nature evolved the appropriate principle of the high-lift, low-drag wing. We may call this the 'lift principle', and nature adhered to it as long as it applied. Below a certain level of size this principle became untenable, and then nature had to adapt itself to the new physical facts. When some insects came down to pigmy size the opposite principle applied, giving a high drag, accompanied by small lift: what we may call the 'drag principle' wing. Thus nature made the best of things. Instead of continuing to struggle against drag, the smallest insect fliers actually enlisted its aid to produce lifting force!

Obviously this could not be done with the existing type of wing, which was designed for precisely the opposite situation. It was necessary to design a completely new organ of flight, and this took the form of the miniature specialist wing the blade of which was composed of numerous parallel bristly hairs. Since the physical problems arising from small size are the same for insects of all the orders, these specialist wings all have a similar look about them. They have reached this resemblance by convergent evolution.

How, then, does this flight mechanism work on the 'drag principle'? To understand it we must take advantage of another kind of convergence. Look at Fig. 38, but for the moment without reading the caption. What do you see there? A drawing taken from a photomicrograph of a bristly wing, showing clearly the club-shaped stalk and the fan-like fringe of bristly hairs. In fact this drawing is of something quite different! It is the hind leg of the water-beetle *Acilius,* a smaller relative of the giant water-beetle *(Dytiscus). Acilius* swims under water, and in the process uses its hind legs like oars to propel itself through the water.

Whether it is a bristly wing or the hind leg of a water beetle, every oar-like appendage operates on the drag principle. Consider how a boat is rowed. The oarsman lowers his blade into the water and pulls against it, and the first reaction he feels is the drag, or resistance, of the water against the blade. Just because the blade is held back, the oarsman can use his oar to propel the boat in the right direction, and the more resistance the blade meets within the water, the more acceleration can be given to the boat. Ideally the blade should not move through the water at all, but be held immobile from the moment that it is immersed. The resistance or drag would then be maximal. Each stroke would take place without any 'slip', and would move the boat the maximum possible distance through the water. It would simply be a matter of pulling hard enough. Of course it is never true that the blade is held quite immobile in the water, but any desired level of resistance can be attained by making the blade bigger and broader. In general, the bigger the blade, the better; and the smaller it is, the less effective. Without a blade on the oar, using nothing but the round shaft, only a very small resistance would be felt, and hardly any forward motion could be produced.

The only other device needed is some mechanism by which the resistance can be reduced during the forward stroke. In a rowing boat the oarsman lifts his blade completely out of the water during this period. *Acilius* cannot do this, but achieves the same result by drawing the legs close under its belly as they go forward, and the swim hairs fall back against the stem. On the next power stroke the hairs automatically spread themselves again. Thus the water-beetle creates the necessary oar blade by erecting the broad exceptionally strong swim hairs of its rowing legs, and their automatic erection and collapse is a necessary part of the process. The tiniest flying insects do exactly the same! The hairs along their wings form a blade which 'rows' them through the air.

These minute insects cannot fall to the ground as quickly as can a shot bird, or a mayfly dying after its nuptial flight. Because of their small size and very small weight, they fall very, very slowly, somewhat like a small coin dropped into a pot of honey. Indeed to tiny insects the air is as viscous as honey is to a coin. To them the air is quite a different medium from the one we know, and they can propel themselves through it with much less trouble than we can, rowing themselves as an oarsman rows his boat, and as an *Acilius* swims through the water with its oarlike legs. All the factors concerned, the size and velocity of the moving bodies, and the density and viscosity of the surrounding medium, can be summarised by saying that for the tiniest insects the air behaves like a thick syrup, through which they can fly in the same way that a water-beetle can swim through water.

So far no-one has been able to make direct measurements of wing movements on this diminutive scale, but nevertheless it can be asserted without much fear of contradiction that things happen as we have described, because the physical laws make them unavoidable.

A single direct observation is relevant here. There are insects that actually 'fly' under water! Tiny Ichneumonidae of the genus *Agriotypus* plunge from the air into the water, and simply continue to 'fly' there with the use of their two pairs of wings. Admittedly the frequency of wing beat is considerably slowed as a result of the very much denser medium, and becomes only two strokes per second, but there is absolutely no reason to suppose that *Agriotypus* uses its wings under water in any way differently from the way in which it uses them in the air. It 'rows' through the water just as it must 'row' through the air, though the latter progress is, of course, much faster. These tiny parasitic Hymenoptera go into the water mainly in search of caddisfly larvae which they pierce with their ovipositor in order to lay an egg in them. The host larva dies after pupating, and later the adult *Agriotypus* emerges from the dead pupa of the host.

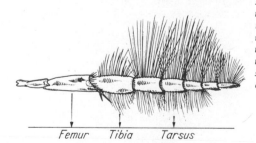

Fig. 38. Right hind leg of the water beetle Acilus sulcatus L. (Dytiscidae), *a smaller version of the well-known great water beetle,* Dytiscus. *The leg is shown in the rowing attitude. The flattened segments present their flat side to the direction of stroke, and the array of strong swim-hairs on the tibia as well as the five tarsal segments is spread out to its maximum extent. The functional resemblance between this swimming leg and the bristly wings of some insects is immediately apparent.*

Femur   Tibia   Tarsus

27. Hoverflies, the helicopters of the animal kingdom

Anyone who has stood near an ivy-covered wall on a day in late autumn, and seen numbers of bees buzzing over it must have asked himself what are they doing there. The loud hum of hundreds of pairs of wings is quite unusual for this time of year. Furthermore, these are not bees, but hoverflies, which look like bees, and which can hover over the ivy, motionless in the air, without effort. Attractive insects such as these can be seen along a woodland path as spots of light, sparkling in the rays of light which filter down through the branches, slowly rising and falling, then hanging poised, completely motionless for several seconds at a time. Apart from the beating of the wings, the only movement is a turning of the head from side-to-side, and a slight inclination of the body to correspond. After a little while the fly suddenly darts away, more quickly than the eye can follow, only to reappear a moment later, hovering poised in a different sunbeam. Hoverflies often have a conspicuous pattern of yellow and black bands on the flattened upper surface of the abdomen. Plate 13 shows two photographs of one of these aerial acrobats. Others keep their aerial watch up in the space between the leaves at the top of a beech tree, high above the ground, and from below they are seen against the light, with a shining yellow spot on the abdomen.

We spoke earlier about ovipositing beeflies (Bombyliidae) which hover expectantly like little bombers, ready to drop their load into the burrows of ground-living bees. These, too, are expert hoverers, though they are not, strictly speaking, 'hoverflies' (Syrphidae).

If you lie lazily under a hedgerow with arms outstretched one of these tiny helicopters will be sure to arrive and take up station, first over one finger-tip, then over another, and occasionally hanging motionless in the air a centimetre or two above the skin, as if it were transfixed. Make a tiny movement, and—flash—it is gone, like lightning, too quickly to be seen. Soon it is back again on the other arm, and begins its tour of inspection all over again.

Hoverflies are the best helicopters in the animal kingdom. Other insects can hover stationary in the air for short periods, the big dragonflies doing this particularly well. When the privet hawk moth, or the jolly little hummingbird hawk moth is sucking nectar it hovers motionless in front of a blossom and probes with its tiny proboscis like an aircraft refuelling in flight. Hawks, owls, and many other birds, including even sparrows, are able to remain poised on one spot for several seconds, but only hoverflies are capable of sustained hovering effortlessly, and with such exquisite control.

The aerodynamics of hovering have already been described in Chapter 15 and so need not be repeated here. It is only necessary to understand clearly that the hoverer's body is tilted backwards so that the wings beat, not at an angle, but in a horizontal plane, functioning like a horizontal extractor fan. They draw in air from directly above and expel it straight downwards. This results in an upwardly directed force on the body of the hoverer, and as soon as this is equal to the weight the body is held poised on one spot.

Any fly can be used to demonstrate the transition from airliner to helicopter if the room lighting is turned off and only a small red light is left burning for the observer to see by. Since the fly cannot see red light, the room suddenly seems dark to it, and it cautiously reduces speed to a progressively slower and slower flight. As soon as the fly reaches a wall it alights, and usually does not fly any further. When hoverflies come to rest in the air, they make a similar change of attitude of the body, though not such a big one. Their whole structure is fitted for hovering in one spot as part of the biological adaptation of these particular insects. They not only buzz when they must, as blowflies do, but keep it up by the hour all day long. There is nothing known to man that nature has not also attempted, and these smallest of aircraft are no exception. Hoverflies are nature's contribution to helicopter construction. Can we learn from their technology?

In our observations on the flight motor we considered the muscle simply as something which can contract and produce force. At that time we were not concerned with how the muscle works; we were only studying how the motor controls the wings. Now let us look at things from the other end. We shall simply assume that the motor is a good motor and proceed to investigate how the central nervous system controls the muscles themselves.

To put it in technical terminology, the muscles are the pistons and cylinders of a flight motor. One may investigate the way in which these pistons turn the crankshaft, so that force is applied to the gear train which finally turns the propeller. That is what we did in Chapter 18. But one can also begin at the very beginning and ask how the individual cylinders are synchronised, when the fuel is injected, when the valves move, at what point the spark plugs ignite the mixture, and how the distributor manages to co-ordinate the firing of the spark plugs. The production of force in the cylinder stands in the middle of this sequence of events. One can observe what is eventually done with the force (cylinder→ propeller, or muscle → wing). But one can also ask how this regularly repeated production of force is controlled (distributor → cylinder, or CNS → muscle). And that is what we want to look at briefly now.

The most important thing is the rhythm. When do the individual muscles contract during the course of a wing beat? To answer this question, we first have to find a method for determining just when a muscle contraction occurs. After all, there are many different muscles in the flight motor; from outside it is impossible to see what each one is doing at any moment. If one cuts the thorax open at the side one can see the muscles, but then the motor will not longer work. Even if the motor should survive this operation, not much would have been gained. Everything would be going on much too fast, and would give an impression of wild confusion. What then—give up? No, there is a very nice thing about muscles— they have action potentials.

Let's start with an example. Some blocks of flats have automatic openers for the door to the building; when someone in one of the flats pushes a button, an electromagnet pulls back the latch and the outside door opens. What has happened? He has closed an electrical circuit, allowing current to flow through the coils of the magnet and generate a field which pulls on the iron core of the opening mechanism.

Another example is this: the large hand of the railway-station clock jumps forward once each minute. In many types of clock that happens because a master 'central clock' (see the functional similarity with the expression 'central nervous system') has decided that all of the station clocks under its control shall advance simultaneously. It closes a contact for a fraction of a second, and this sends out a pulse of current to all the secondary clocks. This is all the central clock has to do. It need not concern itself with actually turning all the hands of the other clocks; it must only give the command with its impulse. Each secondary clock has a built-in electrical ratchet mechanism, which is usually doing nothing. Whenever an electrical pulse arrives a relay closes briefly and the mechanism begins to work. The impulse doesn't actually drive the mechanism, it simply gives a command by way of the relay that the switch should now go into action. And the mechanism follows the command. From its *own electrical supply* it draws energy and heats up a filament in a small cylinder; the air expands, pushes out a piston, and thus sets in motion a simple gear-work which moves the clock hand forward by one minute. As soon as that has been done, the ratchet mechanism automatically shuts itself off. Everything is quiet for another minute, until a new impulse comes from the main clock which gives the accessory clocks the command to do another two seconds' work. This pattern repeats itself every minute.

It is important that controlling the accessory clocks is a completely separate matter from providing them with energy. Each clock has its own movement. The main clock only telephones now and then to say that it is time for the movement to be activated briefly. The main clock controls, but it does not do any of the actual labour.

The distributor in a combustion engine functions in a similar way. At the right moment it gives the command 'now set loose some energy!' This command takes the form of a short high-voltage impulse which makes a spark at the plugs. The combustible mixture follows the command and explodes, moving the piston and doing work. Here, too, control and labour are completely separated. The only thing new, in comparison with the clock, is that the motor itself determines the ignition rhythm, because it turns the rotor in the distributor. The system only needs a little push at the beginning, an electrical starting procedure, and from then on it can run by itself: impulse to the spark plugs—resulting in ignition—resulting in movement of the pistons—resulting in turning of the rotor—resulting in closing of the next contact—resulting in another impulse to the spark plugs, and so on . . .

The station clock is said to be externally controlled (it has no influence upon the signal from the main clock), while the motor is self-controlled (it determines when the next signal for ignition shall be given). But in both cases the command system (main clock; ignition) is completely separated from the system doing the work (the movements of the subordinate clocks, the cylinder and piston). Some are masters, and others are slaves.

It is just the same with the muscle motor: there is a master—the CNS—and there are slaves—the muscle fibres. From the CNS is sent out a nerve impulse, or sometimes a whole burst of impulses at once, to the muscle. The nerve impulse, or spike, carries the message 'do work!' This causes the muscle fibre to turn on its chemical machinery, provide itself with some energy, and to convert this energy into the work of contraction, thus the fibre shortens briefly. Then it turns off the machinery and stretches out again. Now the muscle fibre waits for a new impulse from the CNS. As long as none comes, it does no work. Its machinery is at rest. So there is in principle no difference between this control system and that of the station clocks—the two systems are completely analogous.

We now know that the muscle is a little machine which works briefly on command, and that the CNS is a control centre which distributes these commands. It happens that electrical impulses have a wonderful property: they are easy to measure. If you can detect a nerve spike just before it gets to the muscle, then you can be sure that in the next instant the muscle is going to contract! But there is better yet to come. In the short period when the chemical machinery of the muscle is running it produces, as a sort of by-product, a voltage of its own: the muscle action potential.

Our initial problem was to pinpoint the moment when the muscle contracts. At first the problem seemed insoluble. But now the whole thing looks ridiculously easy. One sticks a wire into the muscle and connects it with a voltmeter to measure the potentials. When the muscle twitches, the action potential it develops is seen as a voltage which moves the pointer of the voltmeter. So we know that whenever the pointer moves, the muscle is contracting. A simple and elegant method since one need not seek out the extremely delicate nerves at all, but can stick the electrode directly into the comparatively thick muscle. This is simpler and gives the same result, since a nerve impulse induces a muscle impulse.

We won't go into the details of how the technical difficulties were overcome, but just say here that it has been possible to solve them. The weak action potentials (a few thousandths

to hundredths of a volt) must first be amplified, so one needs special pre-amplifiers. Instead of instruments with pointers, nowadays one uses inertia-free cathode-ray oscilloscopes. The pictures in the next chapter show how such muscle potentials look on an oscilloscope screen. To record the picture on the screen one needs special cameras. The collection of electronic equipment that has to be built up around a grasshopper or fly looks quite imposing. In Berkeley, California, I worked with such an array of apparatus. There, for example, we recorded the action potentials of four muscles simultaneously, so we needed four amplifiers and a four-beam oscilloscope. Since it would be expensive to photograph everything that happened, and since many experiments fail anyway and have to be thrown out, we first recorded the muscle impulses on a four-track magnetic tape. Fortunately rocket technology has required the development of precise tape recorders like the Ampex; we were glad to have one to misuse for our own purposes. We played back the experiments that succeeded from the tape onto the oscilloscope. When something was particularly interesting we took a picture of the oscilloscope screen. Then we could evaluate what was on the film, centimetre by centimetre. The whole set-up cost just about as much as an average house. But the $30,000 spent on equipment returned a profit: in six months we were able, for example, to record on tape more than 200 long-term flights of bluebottle flies under systematically varied conditions. We looked at the best experiments on the oscilloscope over and over again—you can play back a tape as often as you like —and only then went on to filming and evaluating the results.

One thing more should be said about the electrodes and the evaluation procedure. Electrodes are the 'wires' that are stuck into the muscle. Sometimes they are really wires, two-hundredths-of-a-millimetre-thick tungsten or platinum threads which are insulated to the tip. They can be sharpened electrolytically down to a thousandth of a millimetre. But sometimes even finer electrodes are necessary. Then one uses small glass tubes which are heated to a red glow and drawn out with a pulling machine until they are so fine that the opening at the tip is only a few ten-thousandths of a millimetre in diameter. These pipettes are filled with an electrically conducting salt solution. With these it is even possible to penetrate individual tiny cells and watch the electrical processes going on in them. The more minute these microelectrodes are, the more enormous the constructions which are built up around them. One needs a baseplate weighing a hundred pounds or more, to damp out any vibrations, and expensive micromanipulators to move the electrode under the microscope the tiniest distances back and forth, up and down, and from side to side. This is a science in itself.

If one wants to evaluate the electrical impulses, to see how the spikes are distributed, what the rules governing co-ordination of the muscles are, and whether perhaps there are hidden rhythms underlying the obvious patterns, one can program an electronic computer according to the requirements and feed in the magnetic tape on which the data are stored. The output device then presents the results directly in the form of diagrams and curves. This is an efficient but rather incongruous thing: a single one of our grasshoppers (costing 10 cents) has often been known to keep the whole computer centre (which cost $5,000,000 to build) busy for hours.

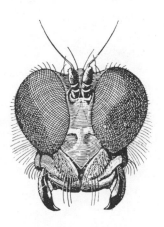

29. The grasshopper beats its wings twenty times per second; the gnat, a thousand times: there are slow and fast flight muscles

When a large butterfly flutters through the air, one doesn't hear it. It is making something like five wing beats in a second (five cycles per second, or hertz), and the human ear cannot perceive such a low frequency as a tone. But we can hear the sonorous buzzing of a may-bug (50 Hz); it sounds like one of the low notes of an organ. Flies buzz in an intermediate range (about 200 Hz). The whine of mosquitoes is at about 500 Hz, and the tiny gnats hold the record with their sharp high tone of over a kilohertz (1000 Hz). The smaller the insect, the higher the flight tone.

We have already learned how well the different flight motors are adapted to these different frequencies. The large insects, such as cockroaches, mayflies or grasshoppers, move their wings slowly and directly. The muscles are accordingly attached directly to the base of the wing. Small insects like beetles, Hymenoptera and flies, as well as many bugs, move their wings very quickly and indirectly by means of the clever click mechanism described in Chapter 19. In this case, the flight muscles do not affect the wings directly, but only cause the thorax to oscillate, which produces a secondary movement of the wings. We can summarise this as follows: large insects have slow wing beats and direct flight motors, small ones have fast wing beats and indirect flight motors.

We have just discovered how one can record muscle twitches in an elegant way, by means of their action potentials. Now it is time to take a closer look at a diagram of what such a recording is like. Fig. 39 shows a reproduction of a four-track magnetic-tape record of an experiment with a desert locust (*Schistocerca*). The details are explained in the legend. It is interesting to compare the phases of the wing oscillation with the time of occurrence of the various impulses. The muscles which raise the wing (the elevators) 'fire'—that is, contract —only at the beginning of the upstroke of the wing and are inactive on the downstroke. The depressor muscles (which pull the wing down) do just the opposite. They fire at the beginning of the downstroke; otherwise they are silent.

This discovery is not particularly amazing. It is just what we had postulated earlier from morphological considerations. But then we couldn't prove it. Now we have proof that it is really so—Fig. 40 shows an original oscilloscope record of locust flight. The situation could have been otherwise, of course. It might have happened that elevator and depressor muscles contracted at all phases of the wing beat, but in different amounts. That would work too, but would be a less efficient use of energy. The records show that this is not what happens. The muscles are still, and thus conserve their energy, in the phases of the oscillation in which they are not absolutely needed.

In Fig. 41 you see a quite similar record, again a photograph of an oscilloscope screen, of three different depressor muscles on the left and right sides. The animal is flying straight ahead. This confirms that all muscles fire at exactly the same time. If they did not do this, the animal could not fly straight.

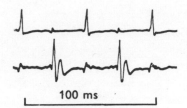

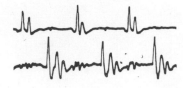

100 ms

Fig. 40. Original recording of the flight of a desert locust (Schistocerca gregaria). In each of the two dual-trace records the upper trace is from a direct wing-lifting muscle (in the metathorax), the lower from a direct wing-lowering muscle (in the mesothorax). The upper record shows very nicely that the schematic drawing in Fig. 39 is correct: elevator and depressor fire with exactly the same rhythm of about 20 Hz, as the time mark indicates. This is the rhythm of the wing beat, so there is a strict 1:1 correlation. Furthermore, they fire in exact alternation, the depressor precisely in the middle of the elevator period (small artefact on the elevator baseline; see the legend of Fig. 42) and vice versa. It is interesting that the elevator fires only one spike, while the depressor discharges a burst of two spikes. The lower record was made somewhat later in the same flight. Here the elevator shows an impulse burst with two discharges and the depressor, one with three; furthermore, the interval from one burst to the next has become smaller, i.e. the wingbeat frequency has risen. The force exerted has also increased; in free flight the locust would be flying faster or rising more quickly. The impulse bursts can consist of at most four discharges, and the number of discharges can differ among the various muscles at a given time, depending on how the muscles are being used in steering movements, etc.

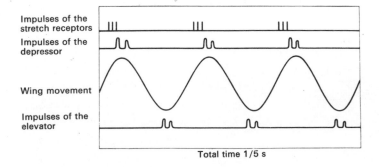

Total time 1/5 s

Fig. 39. Diagram describing the temporal sequence of the impulses in the stretch receptors and the elevator and depressor muscles compared with the wing movement in a locust.

The final results for the locust then are these: first, each of the different muscles receives its own impulse at just the right point in the wing-oscillation period. With each wing beat the process recurs at precisely the same time: the impulses are synchronous with each beat. Therefore one refers to this phenomenon as synchronous excitation of muscles by the nervous system or, since there is usually one impulse per wing beat, as a one-to-one correlation. The second finding is that elevator and depressor muscles are excited in alternation by the CNS, always in the correct phase of a beat period. That is, the CNS not only sends out one impulse or impulse burst per wing beat, but it also makes certain to send it at just the right moment, when the wing is in the precisely appropriate position. This is true for every muscle. And since one muscle must twitch during the upstroke, another during the downstroke, one may speak of a phasic control of the flight motor by the central nervous system. When the rhythm of the wings is controlled by the nervous system in this way, it is called a 'neurogenic' rhythm.

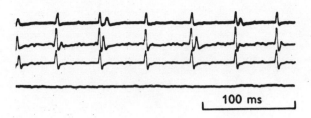

Fig. 41. Another original record of the flight of the desert locust, Schistocerca gregaria. *The three upper lines show discharges from three different depressor muscles of the metathorax. This record shows very nicely how exactly the contractions of functionally similar muscles are synchronised. Sometimes a muscle tends to fire twice (see Fig. 40). The bottom trace is from an electrical torque-indicator which measures the tendency of the animal to roll. It shows no deviation, that is, the animal has no turning tendency and is set for straight flight.*

Here is a final summary of what we know about the flight system of the locust:

1. Large animals
2. Long wings
3. Low wing-beat frequency (20 Hz maximum)
4. Direct flight motor
5. Simple joint system; no click mechanism
6. Synchronous muscle excitation
7. One-to-one correlation
8. Phasic control
9. Neurogenic rhythm.

If we compare these nine points with one another in all possible combinations and reflect upon them for a while, we shall see that all these things are related to each other in a wonderfully sensible way. There could not be one without

Fig. 42. Muscle action potentials in the left (upper curve) and corresponding right (lower curve) dorsoventral muscle IV of a calliphorid fly. The upper pair of curves was recorded during vigorous flight. The upward force, or 'lift', was three times the body weight. The lower pair of curves was made during normal exertion of energy; the lift was just equal to the body weight. It is obvious that the frequency in the latter case is significantly smaller. The voltage fluctuations were played back from tape onto the oscilloscope screen and photographed from there. Since the two beams were sloppily positioned with respect to one another, the two curves overlap occasionally. The very small rhythmic oscillations in the baseline of the upper curve are artefacts, i.e. disturbances, which were induced by the wing beat. Each lasts approximately 1/180 of a second. One can see that between two potential peaks there are roughly 10–15 wing beats, sometimes more, sometimes less; the irregular discharge pattern which is typical for flies is clear. Further, one can see that the phase position

(for example, the distance of a large spike in the lower trace from that 'associated' with it in the upper, as a per cent of the whole period) is also irregular. In the upper curve the first four pairs of spikes occur at a more or less constant interval with respect to one another, but with the fifth and sixth the lower spike suddenly jumps into about the middle of the period of the upper. The smaller, upwards-pointing voltage peaks in each trace come from another unit in the same muscle, further away from the tip of the electrode. These also show the irregular phase relation of 'large spike' to 'small spike' within each trace. The downward-pointing notches, on the other hand, are also artefacts which represent stray voltage or 'crosstalk' from the other electrode. That is easy to tell because they are of the reverse polarity (pointing down) and in contrast to the true spikes they are exactly correlated in time with large potentials in the other trace. This can be seen especially well where the lines overlap a bit; every potential in the lower line corresponds to a small artefact in the upper, and vice versa.

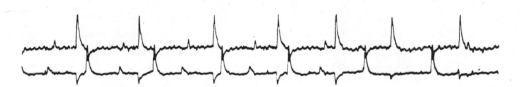

Lift = 3 × body weight

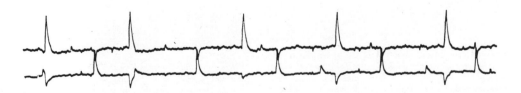

Lift = body weight

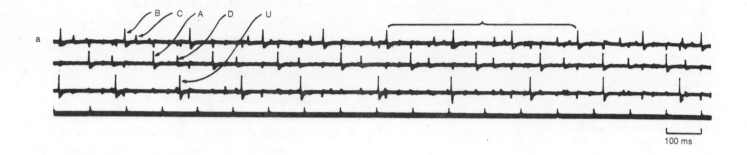

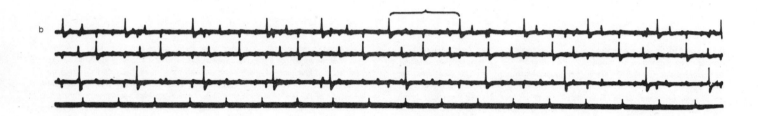

the other, and everything which follows depends upon what has gone before. Altogether it amounts to a marvellously functioning type of insect flight motor. We shall name this type after one of the nine points, let us say point 6, and call it the 'synchronous type'.

Since everything generally has an opposite, we may suspect that there is also such a thing as a non-synchronous, or 'asynchronous' type of motor. And that brings us to the flies.

Fig. 42 shows a record which is similar in principle to Fig. 41, but this time comes from the bluebottle, *Calliphora erythrocephala*. The activity of the left indirect dorsoventral muscle bundle is shown next to that of the right one. How different the details of this record are from those in the locust record! Not only is the wing frequency much higher, not only are the muscle potentials larger and longer lasting; the most important finding, which is a bit difficult to see over such a short period of time, is that the muscles which raise and lower the wings can fire at any time. There is absolutely no correlation with the phase of the wing beat!

The depressors can just as well fire during the upstroke or at its peak or anywhere at all. When one compares many spikes and analyses them statistically, one finds that each muscle is equally likely to fire at any point in the wing-beat cycle, that is, when the wing is in any position. That means that spike occurrence and wing beat are not synchronised.

Quite the contrary—it is obvious that the potentials do not even occur together with every single wing beat, but with every fifteenth wing beat or so. Furthermore, the intervals between spikes are not as regular as we found in the locust; one spike may follow another a little earlier or later than expected. So nothing is fixed at all, and when we compare the elevator and depressor muscles (Fig. 43) as we did for the locust, we find the same chaotic picture.

Our final results: each muscle does receive an impulse now and then, but that is about all we can say. There is no correlation with number or phase of the wing beat, and the relation between the spikes in the two muscles is constantly changing.

Again let us summarise what we know; in order to make the list complete, we mention in point 9 something that we shall not discuss until later: the rhythm is not neurogenic, but myogenic. That is, it is not controlled by the CNS, but arises within the muscles themselves! The list looks like this:

1. Small animals
2. Short wings
3. High wing-beat frequency (150 to over 1000 Hz)
4. Indirect flight motor
5. Bistable joint system with click mechanism
6. Asynchronous muscle excitation
7. No 1:1 correlation, nor any other kind
8. No phasic control, nor any other kind
9. Myogenic rhythm

So we call this type, again after point 6, the 'asynchronous type'.

With this peculiar type of motor it is not so easy to see the relations between the elements. Points 1 to 5 fit together splendidly, but what about points 6 to 8?

We can understand them best if we repeat what we found out earlier, that the nervous system cannot keep up at these

high wing-beat frequencies. That is, it can no longer fire once for each wing beat nor is it capable of always firing at just the right moment. The oscillation period of the locust was relatively long and the spike was short. That meant that the spike could be positioned exactly at any desired point in the oscillation. That is now out of the question: with the high wing-beat frequencies of the small insects the spike gradually becomes longer than the whole period! Therefore synchronous excitation must give way to asynchronous, phasic control to non-phasic, and the 1:1 correlation must be replaced by a variable 10:1 or even 20:1 correlation. And under these circumstances it is necessary to build a new motor which is not dependent on exact temporal regulation. It must be a new kind of motor in principle; the methods used in the locust will not work any more.

The two things most difficult to understand are probably: how can the wing oscillate properly if the impulses can come at the wrong time (for example, a downstroke impulse when the wing is in its upstroke), and how does it happen that the whole mechanism can go on oscillating by itself ten or twenty times between one impulse and the next, rather than coming to a stop?

There is only one logically satisfying answer: control of the wing oscillation doesn't come from the CNS impulses at all! This makes things very confusing. What are the impulses from the CNS for, then? And what does control the wing oscillation?

Imagine an old grandfather clock with a broken movement. If you're lucky, you can still make it run with a little work. All you have to do is stand near the pendulum and give it a little push each time it begins to swing away from you. That way you are providing the necessary motive force synchronously, in a one-to-one relation to the pendulum oscillation, and phasically (always at the moment it begins to swing away) just as was the case for the perfect locust-type motor. This method of keeping the clock going is rather tiresome since you must constantly make sure that the frequency and phase of your hand movements are correct.

Soon you decide you've had enough of this, and you send the clock to be repaired. Now you still have to provide the motive power, but simply by winding the clock once a day; that is, you use the force of your muscles to lift a weight which provides a certain reserve of energy. It is only when this reserve is empty, when the weight touches the ground, that you have to raise it again. But you could do this an hour or even half a day earlier, just as you like. You are now providing the motive force asynchronously, in something like a 20,000:1 relationship, and completely non-phasically. Furthermore, winding the clock now takes longer than one pendulum oscillation.

This is how we imagine the perfect motor of the fly type. And thereby we answer both of the questions posed above. If an energy reservoir is available, and some smoothly operating machinery, then (*a*) it makes no difference at what particular moment one winds it up, and (*b*) one can wait for the next filling of the reservoir until some 10,000 strokes have been made.

The machinery of the fly's thorax operates with this sort of energetic efficiency. The muscles and the click mechanism together form a system in themselves which can oscillate for a long time after it has been given a starting push. This is because the indirect muscles of the flies have a highly unusual property: they do not contract in response to the impulse from the CNS, but rather when they have been given a short, sharp pull! This is so peculiar, and such a departure from the operation of a 'normal' muscle, that for a long time it wasn't believed. The phenomenon is known as a myogenic rhythm (Greek *myon,* a muscle) since it arises in the muscle itself which twitches only when it has been mechanically stretched.

This is how the flight motor of the fly works: the longitudinal muscles contract, this sets the wing into downward motion, the click mechanism is put into tension, and the dorsoventral muscles, which are laid out in the thorax in the opposite direction, are stretched. Suddenly the click mechanism snaps over. This not only accelerates the downward motion of the wing, but it also gives a powerful jerk to the whole structure of the thorax. This jerk is transmitted to the pre-stretched dorsoventral muscles. It is this jerk—and not a nerve impulse—which is their command to contract. They do so and begin to lift the wing, straining the click mechanism in the other direction and stretching the longitudinal muscles. As soon as the click structure snaps back into the first position there is another jerk in the thorax. This time it is the longitudinal muscles which are pre-stretched, and the jerk is a signal for them to contract. The process starts again from the beginning.

And it would go on for ever like that, back and forth, if it were not for energy losses: friction, heat and air resistance to the wings. After some time the system has to be 'wound up' again. That is taken care of by the next impulse from the CNS. It commands the chemical machinery in the muscle fibres to shift into high gear and put out a new surge of energy. But it has no influence on how this energy is spent; that is left to the collaboration of click mechanism and muscle

twitches. After some ten to twenty oscillations the energy reserve has again become dangerously low, but by then another 'telephone call' comes from the CNS and arranges for a fresh supply. And it is not important whether it does this a bit sooner or later. It is irrelevant at which phase of the beat the call comes: whether the various muscles receive their impulses simultaneously or one after the other is quite immaterial. The only important thing is that the new supply of energy should be ordered before the reservoir is empty. But it does not matter at all at what specific moment the refill happens to be delivered.

Let us close with the technical analogy we started with. The locust runs like a system of railway-station clocks; its motor is externally controlled. The CNS determines just when each muscle is supposed to twitch. It controls the ignition rhythm, and in addition it makes certain that with each ignition some energy is mobilised at the proper moment. The CNS works as an ignition distributor and injection pump in one.

The fly, on the other hand, runs like a two-cylinder combustion engine. The motor controls itself. The first cylinder (the longitudinal muscles) controls the ignition time for the second (dorsoventral muscles), and vice versa, just as a well-synchronised distributor does. The CNS doesn't bother about the ignition rhythm, but it makes sure that new supplies of energy are provided from time to time. In flies, the CNS works neither as distributor nor as injection pump, but like the float in a carburettor, which causes new fuel to flow in when the amount given previously has been used up.

So that is the new development in the flies. Smaller animals must have higher wing beat frequencies for physical reasons. But in the CNS there is an upper limit on impulse frequency for physiological reasons. Therefore the locust type could not in the course of evolution, which produced smaller animals, simply be made smaller. Something novel in principle had to be invented, and it was this interplay of click mechanism, muscle specialisation and another kind of control by the CNS. Again everything seems to have been forced to turn out as it has, just as in the development of bristle wings in the Lilliputians. Perhaps with this description I've been able to give you a little of the feeling of being an explorer which sometimes overcomes the biologist. Particularly when, like a stroke of lightning, he suddenly feels that he has perceived a principle of organisation hidden in a mountain of apparently unconnected data.

Viewed in this way—as I hope—even preoccupation with the thorax of a fly may seem to the reader to have a certain justification. Even we biologists, who believe that we understand the details, every now and then shake our heads in astonishment. A little blue buzzer like that—what a subtly clever piece of construction it is!

A couple of dozen hexagonal pencils pressed into a bundle and rolled up firmly in wrapping paper, provides a rather good model of a thin muscle bundle. The wrapping paper represents the outer sheath of the muscle and each pencil, a muscle fibre. In thicker muscles there are inner sheaths which gather single fibre bundles into larger groups. The details of this fibre arrangement are not of interest here. The important thing is that one such muscle fibre represents a single stretched-out and highly specialised cell. It is the functional unit of the whole muscle bundle. It always contains many nuclei, which is a sign that during its embryological development many cells have fused together. Therefore it is common to call this a cell syncytium rather than a typical single cell.

If one cuts across the bundle of pencils and looks at the cut surface or if one simply looks at the blunt end of the bundle, one sees a two-dimensional hexagonal honeycomb structure. Since regular hexagons can be stacked up without leaving any holes between, there is essentially no empty space between one pencil and its six neighbours; the area is used up as efficiently as possible. The cross-section through a muscle shows a basically similar picture, though the hexagons are not formed absolutely regularly. In Plate 14 there is a photograph of a small section of the flight muscle of one kind of dragonfly, taken with an electron microscope, and enlarged approximately four thousand times. You see a cross-section of one 'pencil'—actually the end of a cell; the six adjacent neighbours are indicated as well. You are probably astonished by all the things that are packed into this exceedingly delicate thread of a single muscle fibre. The light, radially oriented double bands which form a sort of many-pointed star are the actual contractile elements of this muscle cell. One calls them muscle fibrils or simply fibrils. Their arrangement, in the form of plates with a star-shaped cross-section, is quite typical for the 'normal' insect muscle, found for example in the legs or abdomen. This dragonfly is a primitive insect, and its flight muscles are not yet highly specialised, but rather look very similar to the other muscles of the body. The same thing is true for the other dragonflies. Butterflies, on the other hand, have also already developed cylindrical fibrils, which look round in cross-section.

There is one difference from normal body muscles which even this primitive flight musculature shows. It is generally known that the fibrils can convert delivered energy into the work of contraction—they can twitch if energy is provided—but they cannot produce any energy themselves. The chemical machinery in the muscle, of which we have spoken several times, is located in another structure—the mitochondrion. All the dark bodies wedged in between the rays of the star are mitochondria. The decisive difference between flight- and leg-muscles is that the former have considerably more mitochondria within the fibre than the latter. You can see that hardly any space at all is wasted—mitochondria and fibrils are wedged in together without a gap. In the dragon-flies the round fibrils are actually wrapped up in a mass of mitochondria, just like a heating pipe with insulation.

These mitochondria cannot themselves contract, but rather they deliver the necessary energy to the contractile fibrils. To do this they need, to start with, energy-rich chemical substances; that is, fuel. They find it in the body fluid which surrounds the muscle bundle; from there it diffuses into the muscle fibres. They also require oxygen, which as we know is transported to each single fibre via the air pipes of the tracheal system. The finest branches of the tracheae—called tracheoles—do not penetrate through the cell membrane into the single fibres. That is, they do not run directly into the mass of mitochondria. But they stand in intimate contact with the membrane, branching out into very fine endings which push into the cell sheath like the fingers of a glove. The oxygen molecules can diffuse rapidly through the delicate membranes; at the most the distance to the mitochondria is only a few thousandths of a millimetre. The mitochondria do just what a proper petrol engine does: they burn the fuel with the help of oxygen from the air, and thus they produce energy. This energy is transmitted by a motor to the pistons and used to turn the crankshaft. In the flight motor of the insect the energy which is set free is passed to the fibrils and used for contraction of the muscle cell. If all the single fibres do that repeatedly at the same time, the whole muscle contracts rhythmically. It is working like the cylinder and piston of a combustion motor and is thus enabled to do cyclic work.

The energy of the petrol mixture or of the muscle is always freed instantaneously and the moment of liberation must be properly timed. From the distributor there runs an electrical impulse to the cylinder, which generates the ignition spark and causes the mixture to explode. From the CNS there also runs an electrical impulse, along a nerve to the muscle, giving the command that energy should be set free. If this process is to occur instantaneously, all single fibres must receive the command at the same time. A highly-branched 'cable', or nerve net, takes care that the voltage pulse arriving in one single thick cable is led through minute branchings to each single muscle fibre. Many muscles even possess several 'spark-plugs per cylinder'; in that case we speak of a multiple innervation. Several nerve endings run to different points on one and the same fibre. Since the fibres are often very long, the announcement 'contract!' needs some time to pass over the whole length. But if the fibre, for example, is innervated at both ends and the impulses arrive at the same time, only half the time is needed for them to meet at the middle. (Multiple innervation also has a number of other advantages.)

Recently it has even been claimed that there is an excitatory transmission system, the branches of which penetrate deep into the muscle fibre like pine-tree roots into sandy ground. This is called the T-system. A contraction command transmitted over this system might be able to strike simultaneously at many hundreds of places in a fibre. Since the microscopic fibre is comparatively gigantic with respect to the molecular mechanisms of contraction, this system would be of real help in enabling the necessary lightning-quick contraction of the flight muscles. You can see that the complicated submicroscopic structure of the muscle is not lacking in mechanisms all of which are directed toward causing the fibrils to contract at the same time. This contraction is the basis for practically every movement in the animal kingdom. The process is so decisively important that it is worth asking, 'How does a muscle fibril in fact function?'

To answer this we must penetrate still further into the very smallest dimensions, into the world of the molecular construction of biological structures. The photo of Plate 14 was enlarged four thousand times. The electron microscope must enlarge a small part of one single fibril four hundred thousand times before one can recognise any more details. A few square millimetres of the light star-pattern on Plate 14 must be enlarged until they take up an area the size of the whole picture. Then there appears a new, exceedingly fine, marvellous geometrical array; it is portrayed in Plate 14. The enlargement factor is not quite half a million.

One can see a geometrical pattern consisting of large and small dots, which once again are arranged in strictly hexagonal order. Each large dot stands in the centre of a hexagon formed by other large dots. At the same time it is surrounded by a similarly arranged hexagon made up of small dots. Since this again is a cross-section, the dots are only the cut ends of threadlike structures which run lengthwise through the fibril. The thick threads consist of a protein called myosin. The thin threads consist of another protein, actin.

Myosin and actin threads, then, run through the muscle fibril in a strict geometrical arrangement. The dimensions are unimaginably small. The muscle fibre is a few thousandths of a millimetre thick (although there are some, much larger 'giant fibres'); the single fibril is at the most a thousandth of a millimetre thick. The diameter of a myosin thread amounts to only one-hundred-thousandth of a millimetre and that of the actin thread to only half as much. All the long myosin threads give off, in a spiral arrangement, small side branches which together grasp the actin threads like the branches of two closely-neighbouring poplars. The threads do not run through the fibril without interruption; rather, they are composed of separate sections. The myosin and actin sections overlap, as Fig. 44 shows. This causes the fibres to have a striped appearance, a cross-banding which can be easily seen even under the light microscope. According to our current ideas, the contraction of the muscle is produced by lengthwise shifting of the actin and myosin threads so that they interdigitate with each other. So far no-one knows in detail how the whole thing functions. It is possible that the side processes of the myosin threads are movable and by bending draw the two types of threads towards each other.

It is not yet known, then, how the mechanism of contraction works in detail, how the energy provided by the mitochondria is converted into the work of contraction. But it is known that these things happen, and even the energy-rich chemical compound which is synthesised in the chemical factory of the mitochondria, and without which no muscle contraction can proceed, is known. It is adenosine triphosphate, called ATP for short. It is also known that the interaction of the myosin and actin, possibly by way of the side branches, depends upon calcium ions. And finally, the place where the calcium ions are stored and released is known; it is the so-called endoplasmic reticulum, a net-shaped organelle which is wrapped around the individual fibrils.

So this is the picture we have at present of the process of contraction. It applies in principle to all skeletal muscles in the animal kingdom. It is only in the smooth muscles of some vertebrate internal organs or in the highly specialised flight muscles of the flies that departures from this scheme occur.

Let us consider a single muscle fibre from the flight muscles of the dragonfly. A nerve impulse arrives. It is possible that the excitation runs through the channels of the T-system and commands the endoplasmic reticulum to set free its calcium ions all at once. When this happens the calcium ions diffuse into the inter-filament space, where they somehow trigger the interaction of the myosin and actin filaments so that they slide past one another, using the energy of ATP molecules. This change of relative position amounts to contraction. It uses up the chemical fuel; that is, it is split into chemical products which are less rich in energy.

As long as calcium ions are being poured out, ATP continues to be supplied, and as a result the muscle remains in

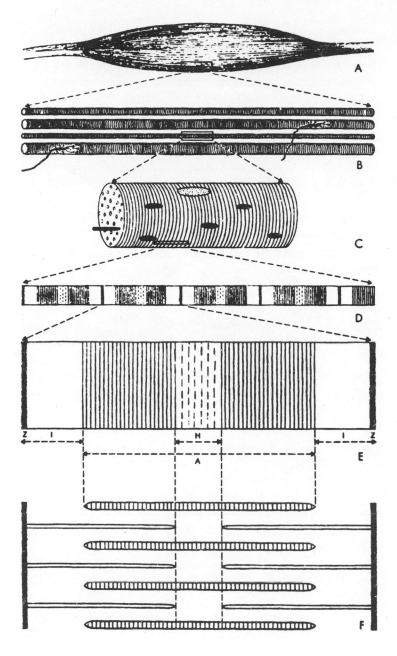

contraction. But that is not what the builder had in mind, since the muscles of the flight motor must of course contract rapidly one after the other and then relax again. This rapid relaxation occurs because the endoplasmic reticulum starts 'recalling' its calcium ions again immediately after it has released them. This causes ATP production in the mitochondria to be stopped immediately, the contraction also stops and the muscle relaxes. Shortly afterwards the next nerve impulse arrives and the process repeats itself, once per wing beat or on the average twenty times a second.

A really accurate description of this course of events would be yet a bit more complicated. The ATP is distributed throughout the myofibril in abundance. When it is required to produce energy it must be broken down into chemical compounds which are less rich in energy. This process of decomposition is controlled by an enzyme.

Enzymes are chemical compounds, often of very complicated structure, which start off and maintain very many biochemical reactions. They are the 'organisers' and 'pacemakers' of the chemical factory.

The enzyme necessary for the splitting of ATP is also present in abundance, but in an inactive state. As soon as the nerve impulse arrives, calcium is released and these ions permit the enzyme to become active, and it only then becomes able to start the process of the chemical decomposition of ATP producing energy. When the calcium ions are captured again, the enzyme again becomes inactive. The splitting process, and with it the provision of energy, is stopped. Interestingly, in this case the enzyme appears to be a combination of the actin and myosin—the very filaments which interdigitate to produce contraction.

Again you see the parallelism with our division of the types of flight muscle. The whole mechanism we have been discussing holds for the flight muscles of the dragonflies, grasshoppers and mayflies; that is, for the slow direct synchronous type. A nerve impulse sets off a cycle; this runs its course, as when a clockwork toy is triggered, and then stops; one-to-one correlation! The next impulse starts the next cycle of complicated activity, and so it goes on.

But flies and bees, and many beetles and bugs, have indirect asynchronous flight muscles without a one-to-one correlation. How does the mechanism function there?

The fine structure of the muscles is in principle the same; only the relative position of the myosin and actin threads looks different. However, there is one remarkable difference from the synchronous type of muscle: the endoplasmic reticulum network is reduced to a tiny remnant. It is possible, then, that these muscles do not operate at all according to the principle of 'control of contraction by release and recapture of calcium'. In any case, they are somehow capable in the presence of calcium and ATP of independent, alternating

Fig. 44. Schematic diagram to clarify the elements of which striated muscle is composed. Each division of the drawing is an enlargement of that part of the division just above it which is designated by the box and dashed arrows. The whole muscle (A) consists of single muscle fibres (B) which appear striped under the microscope, and on which the motor nerves with their 'end-plates' terminate; two nerve endings are drawn. Each muscle fibre consists of myofibrils in a regular parallel arrangement, bordered by nuclei and mitochondria; in (C) one myofibril is sticking out of the cut surface. With greater enlargement (D) the cross-striation is resolved into a regular pattern of units consisting of light and dark bands, and if one further enlarges one such unit one may divide it into clear zones which are designated by the letters Z, I, A and H. Electron-microscopically, each myofibril consists of many, very regularly alternating thick and thin filaments (F) (Plate 15); the ends of these filaments together form the ends of the zones visible under the light microscope which are designated by letters as just described. In the sections (E) and (F) these boundaries are connected by vertical dashed lines. When the filaments slide past each other during contraction, the pattern of bands naturally changes accordingly. This can easily be followed with the light microscope.

movement. For if one dissolves out the endoplasm from muscle cells, bathes what remains in an ATP solution containing calcium ions, and attaches the muscle to an apparatus simulating the wings, it begins of its own accord to make rhythmic oscillations. Under the same conditions grasshopper muscles contract once and then remain contracted; they do not oscillate. This is another indication of the special position occupied by the fly muscles, which seem the more unique the more we know about them.

Today we have the following picture of the way they operate: The physiological releaser for a twitch is not a nerve impulse, but rather a strong abrupt stretch, as we have seen. But the twitch can only occur if calcium ions are present. Nerve impulses make sure that the necessary calcium is set free, and as long as calcium is there the indirect flight muscles, once started, can carry on oscillating rhythmically under continuing intrinsic mechanical control. The greater the expenditure of force demanded of the muscles (more lift!), the more frequently nerve impulses must arrive, so that the 'calcium battery', which is now under a heavier load, can be charged up again more rapidly. This picture is consistent with what we have heard so far about the flight motor of the Diptera. When the nerve impulses stop, the battery very quickly runs down. When this happens the chemical prerequisites for the rhythmic contractions are lacking, and they cease. The fly stops flying.

Finally, we should mention once more that the flight muscles of insects are the most metabolically active tissue that nature has developed. One gram of flight muscle uses more oxygen and fuel, and expends more energy per unit time than a gram of any other tissue. We now understand why: it is because of the exceedingly dense packing of the mitochondria. There are more chemical generators built into the flight muscle than anywhere else. And this is necessary, for in the whole living world no other method of getting from place to place consumes so much energy as the flight of insects. Great expenditure of energy costs a great deal of fuel—from the human point of view, a great deal of money. The person who chooses the most expensive possible means of transportation is permitting himself a luxury. Flying animals are the luxury models of nature.

31. The accelerator pedal for the flight motor: nerve impulses regulate the delivery of muscle energy

If someone wants to catch a bus which he has just seen coming, he must switch from his comfortable walking pace into a brisk trot. From a physical point of view he accelerates his body to a greater velocity and maintains it, in spite of losses owing to friction and the resistance of the air. This requires that more energy be expended than before, so somehow the nervous system must communicate to the muscles that they now must deliver more energy.

When the desert locust wants to fly faster, climb to a higher altitude, or compensate for a strong head- or side-wind, it is faced with the same problem. How does it let its muscles know? At the top of Fig. 40 you see a record you have already seen, made during a quite normal forward flight. Then the record is broken off. Below in that figure you see the same kind of flight record during a sharp climb: here two

or three impulses arrive from the nervous system for each wing beat, and in addition the repetition rate of the impulse bursts increases. The greater expenditure of energy demands stronger excitation. If two nails hold better than one, three should hold really well. But that is about as far as it can go. The locust usually has only a two- or three-gear transmission at its disposal: in the economy gear there is one impulse per wing beat; in ascending gear, when higher performance is required, two or three or at the most, and once in a while, four impulses.

Now, how does the fly do it? In the lower part of Fig. 42 a record is reproduced which was made during normal flight. In this condition, as you know, the lift is equal to the body weight. Above in that figure you see the record of a particularly energy-consuming sharp ascent, during which the lift was just three times the weight. To be sure the potentials still occur just as irregularly as is to be expected with flies, but now they come much more frequently. You know that a muscle action potential is a sign that the muscle motor has been 'wound up' for at least a dozen oscillations. If it is to produce more energy, it must be wound up more often. Therefore the next impulse must come sooner. As long as this state of affairs continues, there come more impulses per second than during normal flight: the discharge frequency rises. That is just what the record shows.

So while the locust has a two- to three-gear transmission, the fly possesses a continuously variable one. The impulse frequency can change smoothly and so does the amount of energy provided by the flight motor. One can measure this energy, for example in terms of the lift force generated by the fly over a certain flight distance. This is shown quite nicely by graphing the data from a film, as is done in Fig. 45.

At the top you see the discharge frequency in a right dorsoventral muscle, in the middle that of a left dorsoventral muscle, and at the bottom the lift is indicated. The data are from the last two minutes of a long flight in which the fly proceeds rather erratically, then changing its lift strongly and finally stops.

You see that all three curves are nicely parallel to one another, like the banks of a river. Since the instrument measuring the lift was oil-damped, it doesn't show very short fluctuations of force. The fact that the impulse frequency in the two muscles changes approximately in parallel is no longer suprising to us—this is the special characteristic of the insect motor. And the fact that the two muscle curves run parallel to the time course of the lift indicates the close and continuous correspondence: more muscle impulses → more fuel → higher flight performance → greater lift → sharp ascent of the animal! The nerve impulses correspond to an accelerator pedal. The more strongly one steps on it, the more fuel can flow and the more force is generated by the motor. With nerve impulses the central nervous system controls and regulates over a continuous range the energy budget of the flight motor. In the last chapter we spoke of the fine structure and the mode of operation of the muscle, and we saw how the energy is delivered and converted in the muscle.

Perhaps you would like to know how we obtained the data in Fig. 45. We let the insect fly in an aerodynamic balance which indicates the lift. Every two seconds the reading at that time was fed to the first channel of the Ampex tape recorder via a microphone. The oscillations of the thorax were transmitted via a fine wire to a tiny pressure transducer fastened to the balance; the signals from this device were automatically fed to the second channel. Two fine platinum wires were introduced as electrodes into the desired part of the musculature. Via these wires the muscle action potentials were fed

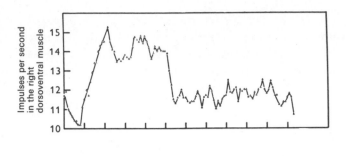

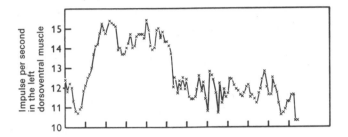

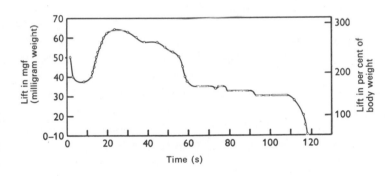

Fig. 45. Evaluation of the last two minutes of a several-minute flight of the blue-bottle fly Calliphora, during which the lift happened to vary strongly, between 300 per cent and 0 per cent of the body weight. One can see that the sequence of impulse frequencies in the right and left dorsoventral muscles is the same and that it also parallels the variation in the lift. The oil-damped balance (see Plate 8) shows the smoothed time-course of the lift; rapid fluctuations are damped out and thus do not appear in the curve of life vs. time.

to the third and fourth channels. The whole procedure of adjusting the electrodes and the pressure transducer could be carried out under a dissection microscope. Then the animal, with all its attached paraphernalia, was fastened to the balance by a tiny, many-pronged plug. Plate 8 shows part of the experimental set-up and a large photograph of the animal shortly before the flight. At this time it was still under the ether anaesthetic and was not yet moving its wings.

If the fly was treated properly, it would fly almost as well as in free flight, but not for as long. The best fly managed three-quarters of an hour without pause. After the first fifty experiments the preparation went very efficiently, with three minutes for the whole procedure. After each individual experiment the fly was killed and dissected under the microscope in order to see whether the tip of the electrode was really stuck into the right muscle.

This dissection always lasted an hour and a half. We had to have the finest possible watchmakers' forceps. A steady hand is needed as well, for one must be able to move the tips of the forceps exactly to within a few hundredths of a millimetre.

32. The steering gear: how does an insect fly right and left?

The last important question which we wish to ask about the flight motor is this. What use is the most beautiful aircraft if it cannot be steered?

It is not difficult to guess how the locust does it. If the nervous system has to tell the direct flight muscles exactly what to do, then presumably it can also issue the commands for steering. This is indeed what happens, and it is done by making slight alterations to the phase of the wing beat at which the nerve impulses are sent. The muscles move each of the four wings more or less independently. If there are differences of timing, then the forces generated by the wings will be different and flight along a straight path will be transformed into flight along a curve. It is also possible that one muscle will receive two impulses per beat, and the other, only one; this too will change the relative forces. In spite of recent research which shows that there may be some separation between the functions of power production and control, it is nonetheless true that in the locust, the muscles which power the flapping of the wings also do the steering. The situation is similar in the dragonfly.

In the flies it is quite different. We have already seen that all the muscles of the flight system receive about the same number of nerve impulses; sometimes many, sometimes only a few, according to the amount of energy needed, but always about the same amount for each muscle. Because each side of the body is the same, the insect cannot generate different forces on both sides by this means.

We can now look at an extreme case. When an insect in free flight turns a sudden corner, or even turns round completely in its tracks—if you try to catch a fly, it often responds with these sharp manoeuvres—it often beats only one of its wings. The other wing is drawn back and held in its rest position. No wonder the animal pivots around under the influence of the one-sided aerodynamic forces. If it makes a sharp turn to the left, it only beats the right wing. One might suppose that action potentials would only be observable in the right-hand side of the insect because the wings on the other side are quite still. In fact each side shows the same number of action potentials. Nothing is changed from the situation in straight flight, so the powerful indirect flight muscles have nothing to do with steering and there must be another group of muscles which perform this function.

There is a group of about a dozen direct flight muscles connected to the base of the wing on each side which have already been mentioned. The musculus latus, involved in the click mechanism, is one of them. These muscles obtain their impulses from another part of the nervous system. The potentials are smaller, shorter and very much more frequent. The CNS may deliver up to 200 spikes per second to one of these tiny control muscles which rotate the wing in the required direction. The wings still beat up and down, but the control muscles don't worry about this. They rotate the

Plate 23. A capture-stroke by the praying mantis, Hierodula, as she hangs upside down. The pictures run from left bottom to right top, with an interval of 10 milliseconds between them. The booty is a black papier mâché disc suspended by a white thread. It is gripped between femur and tibia about 50 milliseconds after the beginning of the stroke, and the retraction of the fore legs is complete somewhere between 100 and 200 milliseconds later. A little rod of balsa wood is attached to the insect's head to make the movement of the head visible.

Plate 24. The house fly, photographed in free flight by means of an electronic flash of 1/5000 s duration. Fly seen from in front (upper photograph) and from above (lower photograph). The attitude of the legs is not typical of free flight, and probably indicates that the fly was photographed shortly before it landed.

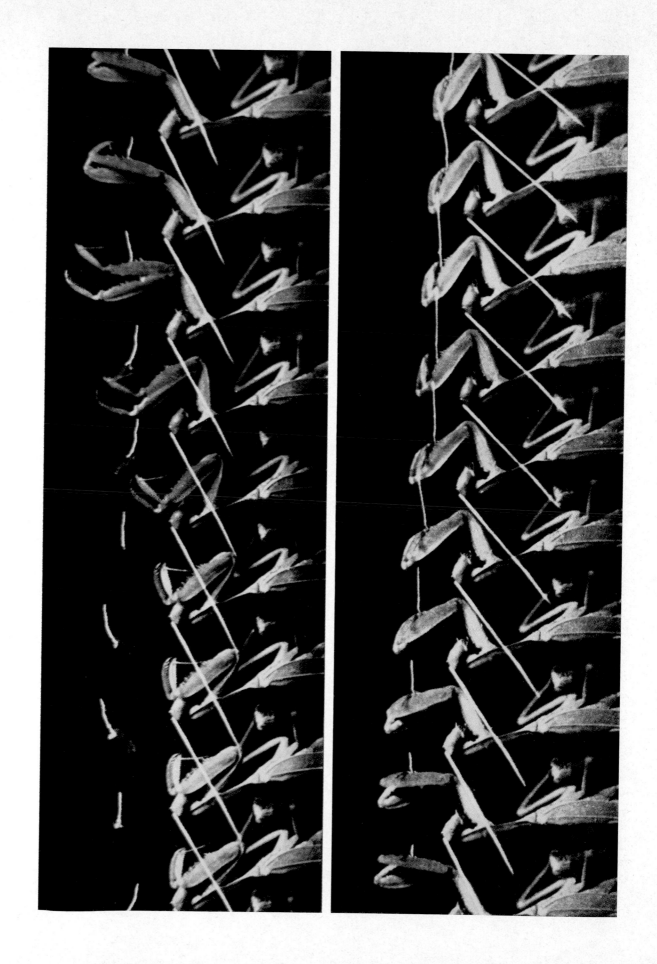

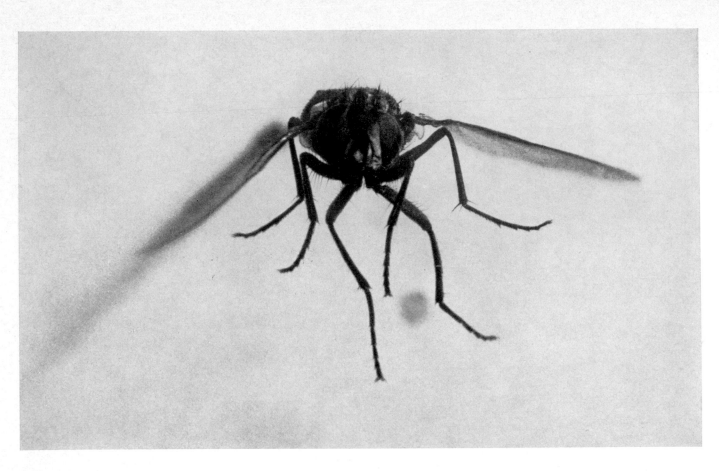

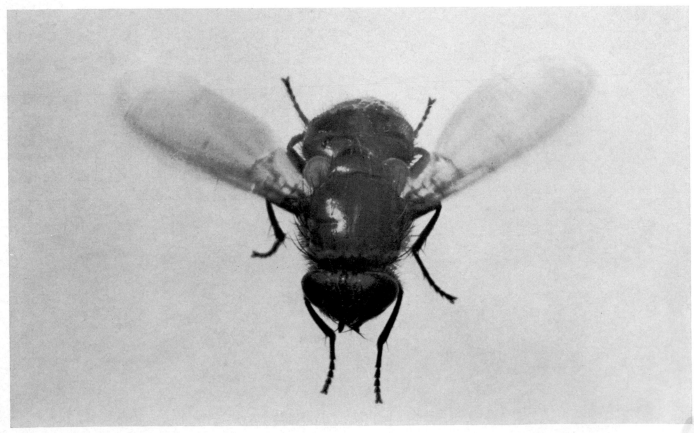

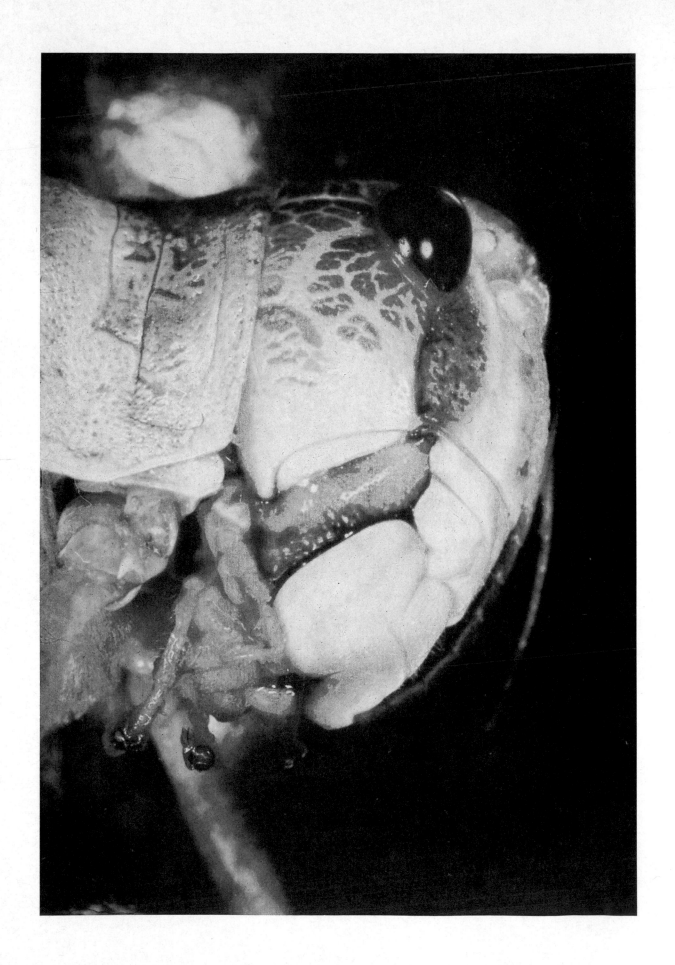

whole wing in the required direction, the right- and left-hand wings independently. It is like controlling two small air-conditioning units, one with each hand. The electric motors in the units produce the power, as the indirect flight muscles do, in causing the fans to rotate. Our arms act like the direct controlling muscles, which can direct the swinging wings in different directions. Our arm muscles are quite indifferent to the fact that the fans are turning, just as the control muscles are indifferent to the fact that the wings are beating up and down. One of these muscles can draw the wing backwards and uncouple it from the still running thoracic machinery (see Plate 16). This is just what happens to the left wing if the insect turns sharply to the left, and similarly to the right wing in a right-hand turn. So the locust has only one muscle system, which at the same time provides both power and control The fly has two systems. One of them produces power but has no steering function; this is the powerful indirect flight musculature. The other controls, but does not contribute to, the power, the small muscles connected directly to the base of the wing. In the flies the driving system and the steering functions are completely separated. It is just the same with aircraft: the engine does the driving and the elevators and the rudder look after the steering.

There is a reason for this complete separation of the power-producing muscles and the control muscles. If two functions are separated, each can be made more suitable for its own purpose, and this can be more highly specialised. Every engineer knows that. And the flies, it is quite clear, are among the most specialised of all flying insects. This is why they are of such interest to biologists, and this is why I went to America in 1966. All because of the muscles of a fly!

Plate 25.   The locust. Taken from the Apocalypse of Saint-Sever, middle eleventh century (perhaps drawn by Stephan Gorcia). Paris, Bibliothèque Nationale. For explanation see text.

Plate 26.   Head of the desert locust Schistocerca gregaria. The broad spoon-shaped structure which encloses the head from below is the upper lip or labrum, and conceals the powerful chewing mandibles.

All insects have six walking legs. Spiders have eight legs, and Crustacea ten including their 'pincers'. Millipedes do not literally have a thousand legs, but anywhere between 30 and 340. Insects are the most specialised class of arthropods, and there is a very sound reason why in the course of evolution they should have reduced the numbers of walking legs to six. This arrangement is well suited to every kind of locomotion.

Consider a table with four, six, eight or even more legs. Whether it stands on an even or an uneven surface it is liable to wobble. A three-legged table will stand firmly no matter how uneven the floor. Tripods are always preferred to tetrapods, whether for photography or as a stand for lunar modules. The six legs of insects function as two tripods. Imagine an insect as seen from above, and number its legs from left to right, and back to front: 1, 2; 3, 4; 5, 6. When the insect walks it lifts three legs—say 1, 4, 5—off the ground, and reaches forward with them while standing firmly on the tripod formed by the other three, 2, 3, 6. On this support it can neither wobble nor fall over. As soon as the first three legs are firmly down again the insect shifts its weight onto this tripod, and, secure and stable on this support, raises the other three legs (2, 3, 6) and moves these forward in turn. Six legs are a particularly suitable number to use in this way, just because three at a time will form a stable support.

During flight the legs are withdrawn, either folded into a compact mass, or at least pressed closely against the under surface of the body. The dust jacket of this book shows a fly with the legs in a typical flight attitude. Only quite slow-flying insects such as craneflies and midges allow their legs to hang down.

The six legs have a special part to play at take-off when they function as a catapult, and on landing when they act as shock-absorbers. Insects do not run forward after landing like an aeroplane, nor do they descend slowly like a heli-copter. Usually they arrive at any angle, alight with a distinct jolt, and stop on the spot. This often calls for a sharp decelera-tion, or braking effort, which must be provided by the legs, outstretched forwards. A cockchafer, which strikes the hard bark of a chestnut tree at a speed of 2 metres per second, and half a centimetre further on has come to rest and remains still, suffers a retardation of 400 metres per second per second, or about 40 g! If the mass of the beetle is reckoned at about 1 g, during such a landing its legs must absorb energy equivalent to two-thousandths of a watt-second. They have about one two-hundredth of a second in which to do this,

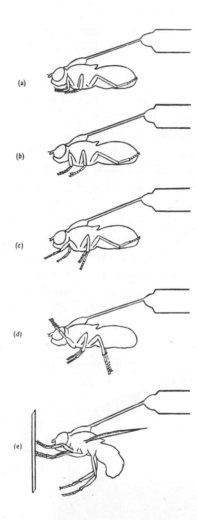

Fig. 46. Flies have a special technique for landing on vertical surfaces, which can be investigated under controlled conditions by fixing a fly—in this instance Lucilia sericata, the common greenbottle fly—by its thorax to a support and moving a small vertical plate towards it mechanically. From the normal flight attitude (a) the fly first of all allows the tarsi of the fore and middle legs to droop slightly (b) and draws the femur and tibia of each of these legs apart from each other (c). Finally, it stretches out the forelegs in front of it, with the tarsi higher than the fly's head (d). Meanwhile the middle legs are gradually inclined forwards, while the hind legs are jerked down-wards out of their flying position. As soon as the fore and middle tarsi touch, the wings abruptly stop beating, and the body is curved downwards until the hind legs also make contact with the alighting surface.

and so must work at a rate of 0.4 watts, or rather more than one two-thousandth of a horsepower. This they cannot do entirely without assistance, so that some part of the shock of landing must be absorbed by elasticity in the sclerites underneath the thorax.

The legs grip the substratum with spines, claws and sticky pads called 'pulvilli', and because of these devices the insect has practically no tendency to roll forwards. A cranefly's legs are widely outstretched when it comes in to land, and as soon as a tarsus of one of the fore legs has made contact the wings stop beating, and the cranefly drops down the few millimetres that are necessary for all the other legs to touch. Dragonflies land with all their legs extended downwards, forming a kind of basket open towards the front (Plate 9). A further use for the same leg-basket is to seize prey in flight. Flies extend all six legs out obliquely in the direction of the landing surface as soon as it comes within a few body-lengths of them. The first pair are held out particularly high. Fig. 46 shows five stages of the approach to land on a vertical wall. The compound eyes alone give the orders for the landing gear to be extended, and so the landing process calls for a very complicated system directed visually.

Incidental note: flies in their abrupt flight must react more quickly and certainly than many other insects, and must obviously be among the most alert of all creatures. They can distinguish precisely between events that are separated by less than one two-hundredth of a second. In this period of time the fly has already travelled forwards its own body length, and so such a speed of reaction is vitally necessary to it. Man's reaction time is ten times slower, and we cannot distinguish events which take place less than one twentieth of a second apart. When we are in the cinema we are not aware that in such a short period of time the picture of the screen has been changed, with a dark interval in between. Our eyes are too slow for that. To a fly, however, a cine film must look like a lantern lecture, with long, dark pauses in between the slides!

34. A catapult launch is nothing remarkable for the fly

When I was working in Berkeley on the flight motor of the Diptera, I also had a quick look at how flies launch themselves. If you have been intrigued by the highly specialised indirect flight muscles you will immediately realise what the main problem is. These muscles do not contract because a nerve impulse has been delivered to them, but because they have been stimulated by being suddenly stretched. This is all very well as long as the flight motor is already working, because then the two pairs of muscles stimulate each other, but which should contract first, which should start the motor? It cannot be one of the indirect muscles, these only function if the system is already running. So there must be a special starter, a muscle which sets the machinery in motion. The muscle must give the thorax a hefty tug, and thereby stretch the indirect muscles. This is the signal for the flight muscles to start contracting. As soon as one of the antagonistic muscles has contracted, it stimulates the other and this, in turn, stimulates the first one, so the process can continue. The same situation occurs in a petrol engine. As long as it is already turning, it can regulate its own action by means of the distributor, but for the first half revolution it needs at least a starter motor. It cannot start itself, however much petrol we pour into it.

As was expected, we often observed in Berkeley that the flight muscles showed a normal pattern of action potentials (which allows the chemical systems to get ready for flight), but the wings of the fly were nevertheless quite still. The fuel was present, but the motor did not start. After much probing around we found the starter muscle. Once more it is a direct flight muscle, one of the most powerful of its type. This muscle runs from the 'hip joint', or coxa, of one of the pairs of legs, to the 'saucepan lid', the scutum. When it contracts, it pulls the lid a short way into the pan, and this, in turn, stretches and stimulates the longitudinal muscles, which begin to contract pulling the wings downwards from their resting position. The motor is now running.

The starter muscle has started the flight musculature mechanically, but itself must be stimulated in the usual way by nerve impulses from the CNS. This must happen as quickly as possible because under some circumstances a lightning start can save the fly's life. It has many enemies who would like to bring it down or capture it, such as predatory flies, Hymenoptera, spiders or praying mantids and the fly may be able to escape from one of these by a quick take-off.

The starter muscle has two special features. The first is that the largest nerve fibres in the whole animal run directly to it from the brain. The larger a fibre is, the faster an impulse travels down it. As soon as the fly has sensed danger and decided to fly away, it can give orders to the starter muscle in the shortest possible time. The second special feature is that the muscle is attached to the leg and not to the nearby wall of the thorax. When it makes a powerful contraction, it not only sets the flight motor in motion, but also automatically forces the middle leg downwards. This happens synchronously on the two sides of the insect. The result is a strong jumping take-off: whilst the flight motor is still in the process of starting, the fly has already projected itself obliquely up into the air, and given its body a considerable acceleration. This is one of the secrets of the lightning launch. It can be demonstrated very easily by blackening a plate with a candle flame, and then allowing a fly to take-off from its surface. The middle legs leave a bright tuft-shaped bare patch, the hind legs a much smaller one and the front legs almost nothing at all. So it is the middle legs which are the catapult or take-off rocket.

This synchronous activity was discovered quite simply by comparing the timing of the action potentials. A muscle which shows an action potential earlier also contracts earlier. We also recorded the sound of the wings, measured the impulse on take-off from a pressure-sensitive platform and observed the fine details of the body movements for a few thousandths of a second before take-off with a weightless light pointer. We discovered that the start, apparently so simple, was composed of a whole series of events which followed one another with intervals of a few thousandths of a second. It is probable that the whole process is somehow automated.

Let us suppose that a praying mantis wants to catch a fly. It remains quite still until the fly has approached near enough, and then, quick as a flash, reaches out with its arms like a pair of pincers. The action is over in a little more than a fiftieth of a second, but let us suppose that it has been extended to over a minute (that is, by a factor of 3000 times) with an imaginary slow-motion cine camera. The compound eye of the fly detects the arms of the mantis as moving objects, and

a computational centre in the brain works out their direction and velocity. The fly prepares for a catapult start in the opposite direction. The brain decides 'start', and sends the order to the starter muscle in the form of one, two or three closely spaced nerve impulses. A few milliseconds later the musculus latus has been given the order 'get the click mechanism ready to go'. It quickly contracts, stiffens the walls of the thorax, and sets up the conditions in which the flight apparatus is able to operate. At the same time the main flight muscles receive the order 'open the fuel valves and leave them open'. The sequence of action potentials in the mighty flight muscles begins but the motor has not yet been started. At this instant the action potentials reach the starter muscle. It contracts violently, the scutum is drawn inwards, the middle legs push powerfully against the ground. The insect leaps from the ground, and the catapult start has begun. At the same time the quick stretch has caused the longitudinal muscles to contract, so that the wings move downwards slightly. The amplitude of the wing beat is very small at first, but quickly increases. While this has been going on, the jump has been propelling the insect obliquely upwards. After six beats, the amplitude of the wing motion reaches its normal high value, and the flight motor is working on full power, accelerating the fly out of reach of the dangerous arms of the mantis. The legs have been retracted to reduce obstructive air resistance.

That was the catapult start, of which the main process, from the stiffening of the thorax to the start of functional wing beats, has taken just twenty-thousandths of a second. The mantis, for its part, must make complicated calculations how to operate its arms, and if they are correct the catapult start is to no avail: the fly is caught. But if the mantis miscalculates, the catapult start gives the fly a good chance of getting away. It is interesting that the capture of the prey takes about the same length of time as the catapult start, taking into account the time taken to recognise the prey and set the body in motion. It too is about 30–50 thousandths of a second. Both parties have a real chance in this deadly but fair battle of a few thousandths of a second. If one of the processes was, say, 20 per cent faster, the other party would have no chance at all. Preying insects have been feeding on flies for millions of years, so right from the very beginning the six hundred or so living species of mantids must have had a capture mechanism which is at least as fast as the catapult start of the fly. That they have succeeded in this is shown by the fact that the capture is successful for more than 85 per cent of the time.

How does the fly manage it? Does it have to fly on its back so that it can reach the ceiling with its feet? Or does it fly just below the ceiling, and then do a half-roll until it can reach the ceiling with the three legs of one side? However much we speculate, it is clear that the operation calls for a complicated piece of flight co-ordination.

In fact the fly carries out this operation in a way that is as simple as it is elegant. The problem was solved by an American research worker, using high speed photography, and I have confirmed it in my own laboratory—it works!

The fly approaches the ceiling obliquely from below, at a steep angle, and at a speed of about 25 centimetres per second. It then flies straight into the ceiling, and shortly before striking it, stretches out all three pairs of legs. The fore pair take up a special attitude, held out stiffly upwards so that they are the first part of the fly's body to make contact. Held in this way they act as shock absorbers and as anchors, adhering at the point of contact by means of their claws and pulvilli (hairy pads). Simultaneously the wings stop beating. Now the fly is clinging firmly to the ceiling with its fore feet, but its body still has a certain forward momentum. Like a flywheel on its shaft the fly rotates about its fore feet and turns its belly upwards, grasping the ceiling with its middle and hind feet—and there it is, sitting upside down on the ceiling, without having had to fly upside down first! All twelve of its pulvilli act like suction cups, and hold it securely. All amazingly simple, yet none of the many people who had speculated about this phenomenon had thought of this possibility.

Fig. 47 shows freehand sketches of this operation as it was revealed by a high-powered stroboscopic light, in the form of silhouettes. The stroboscope was working at a thousand flashes per second, and the time of exposure for a single picture was only one millionth of a second. The film was wound on to a drum, and ran continuously at high speed behind the lens. In spite of this continuous movement of the film, the extremely short individual exposures meant that each picture was 'frozen' as a sharp image on the film.

Fig. 47. Freehand sketches of the landing of a fly on the ceiling drawn from single exposures on stroboscopic film. The successive stages of landing are explained in the text.

This question is not as pointless as it might seem. There are three possibilities: one, the insect may break off its wings and never fly again; two, the wings may be pressed against each other; or three, the hind wings may be folded in a more or less sophisticated way and stowed beneath the fore wings as wingcases.

The first method is mainly confined to ants and termites after their mating flight is over. All four wings fall off together suddenly like a jet-fighter dropping its auxiliary fuel tanks. A termite performs this neatly with a minimum of struggle, so that the wings fall down to the ground in the same relative positions as they occupied when they were attached to the insect's body.

Most butterflies and day-flying moths hold their wings at rest together above the body, exposing the undersides of the wings, which are often very inconspicuously coloured, dull grey or brownish, with irregular patterns predominating and making a form of camouflage. If a peacock butterfly opens its wings in front of a bird the sudden appearance of the upper surfaces, with their brilliantly coloured eyespots may startle the bird and make it fly away. Night-flying moths more often hold the wings inclined together over the body like the roof of a house, or else keep them extended horizontally. In these attitudes the upper surfaces are exposed, and, appropriately, the upper surfaces in these Lepidoptera are the more soberly coloured.

If a noctuid moth settles on a pine trunk it is almost invisible. Locusts and grasshoppers produce a similar effect by folding away their coloured hind wings longitudinally like a fan. Mayflies and damselflies press their wings together upwards, whereas the big dragonflies hold them stretched out horizontally. Flies either lay one wing neatly over the other on top of the abdomen, or hold them obliquely sideways at a slight angle, like a dart. Wasps and hornets fold each wing longitudinally, so that it is much narrower than in the attitude of flight: the German name *Faltenwespen* refers to this. Bees fold their wings at the base in such a way that they can be turned directly backwards, but without altering the breadth of the wing, as shown in Fig. 48. Many tropical cockroaches *roll* their wings longitudinally, like a flag round its staff, a unique example among insects.

The folding methods for beetles are among the best known, especially those of Staphylinidae, which fold their leathery wings lengthwise, transversely and obliquely, into a tiny parcel. Almost every family of beetles has its own way of folding its wings, with ladybirds and burying beetles being specially adept. Many bugs (Hemiptera) have mastered the technique of wing-folding equally well, and the earwigs are unsurpassed at it. Earwigs and beetles are the insect exponents of 'Origami', the Japanese art of paper-folding.

Let us now take a look at one of the most complicated

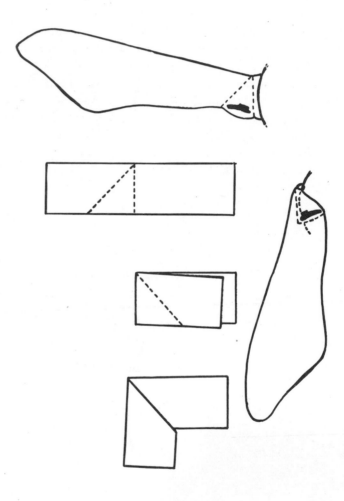

Fig. 48. Folding of the left wing of the honeybee, Apis mellifera, *from the out-stretched position (above) to the resting position (right). To make this clear the diagrams on the left show three stages of folding of a paper model. The first fold is at right angles to the long dimension of the paper, and the second at 45° to this, resulting in the paper being folded into a right angle (lowest left). The black object between the two lines of folding (dotted lines) in the wing drawings is the third axillary sclerite, to which are attached the muscles which fold and extend the wing.*

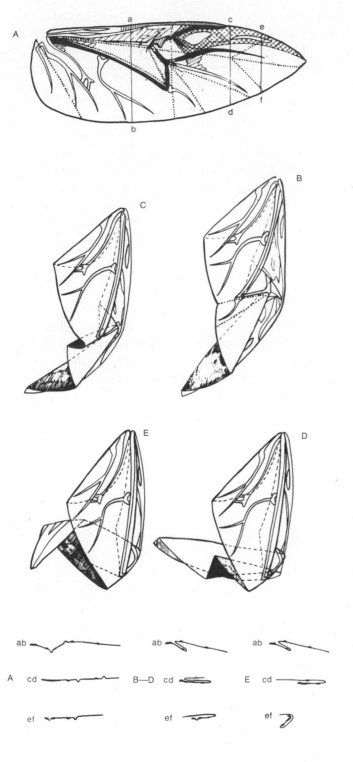

ways of folding the wings among beetles, and learn from it something of the reverse process, the opening of the wings, and their separation from each other. It is difficult to show in a diagram the folding technique of ladybirds, staphylinids and burying beetles, because in their compressed bundles several folds lie superimposed. So let us take a different beetle, which is no less skilful, but which does not bundle up its wings so tightly. I mean the tiger beetle, *Cicindela*. This beetle can often be seen running about in sand pits, that dark-green chap with yellow spots and remarkable mandibles. This is one of the smaller beetles, and does not go in for elaborate preparations for flight. Tiger beetles can extend the wings with lightning speed, and dart away like any fly. This would be impossible if the wings were folded and packed away so elaborately that their unfolding took an appreciable time, but the wings are folded towards each other in two longitudinal folds, and then are turned at right angles in the middle and tucked away under the wingcases. In order that the tip of the wing shall not stick out, it is folded back as a last step. Fig. 49 shows these four stages, step-by-step. An ingenious system of stiffening ridges and interlocking spring mechanisms on the wing membrane keeps the wings folded

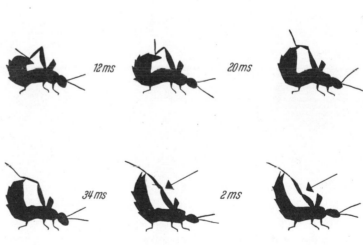

Fig. 50. Diagrams of the process of wing-folding in the small earwig, the native species Labia minor L. The big earwig Forficula folds its wings in an essentially similar way. The sketches are drawn from cine pictures, and the time interval between successive positions shown in the sketches is quoted in milliseconds. In each sketch the left wing is shown precisely in side view, and the artist has cut down the right wing to a mere stump. The process of folding takes a long time—68 milliseconds for the stages shown in this figure, and 1–8 seconds for the complete operation—but it will be seen that the wing 'clicks' from one stable state to another almost instantaneously: 2 milliseconds (1/500 second) is all the time that elapses between the penultimate position (wing still convex in front, see arrow) and the final position (wing clicked back into flying position, see arrow). Plate 3 shows the wing a few milliseconds later.

Fig. 49. Wing folding in the tiger beetle, Cicindela hybrida. The various sclerotised sections in the leading edge of the wing are shown in black, crosshatched or squared. The lines of folding are shown either by broken or dotted lines. At the bottom of the figure are shown cross-sections of the wing taken at three points (ab; cd; ef), and at three stages of the folding process (A, B to D and E). In these cross-sections the veins are indicated by small black swellings on the line. Details are explained in the text.

and pressed against each other, yet springs apart when the wings are opened for flight. Hair tufts on the wings, and hairy ridges on the thorax ensure that the folded wings do not come apart while the insect is running about.

Our native earwig, *Forficula,* has very rarely been seen on the wing, and no wonder: only occasional individuals are even capable of flight, and these rarely fly. The other native genus, *Labia,* is smaller and less well known than *Forficula.* In the resting position the wings of an earwig are closed up like a fan, and then the bundle of wings is folded twice. The 'pincers' that are such a prominent feature of earwigs help in folding the wings, but they are not capable of really acting together like a pair of pincers, and can only make sideways movements. Fig. 50 shows six stages of the unfolding of the wings, by means of silhouettes drawn from frames of a cine film. The left wing is seen exactly end on, while the right wing has been cut down to a stump to avoid confusing the picture. It will be seen how the tip of the abdomen is used to push against the wing, and to smooth out its two folds bit by bit. Notice how this forms a sort of spring trap mechanism. In the penultimate drawing the arrow indicates the point where the wing is still hinged towards the right. This is the first stable position. The final drawing shows this joint now hinged towards the right, the second stable position. The wing is now completely rigid and ready for flight, and of course the right wing has been extended at the same time.

In Plate 2 the upper photograph shows the earwig *Labia* bending its abdomen, and using its pincers to unfold the left wing. The unfolding process is just about to begin. The wing that is already extended clearly shows along its front margin the dark, jointed, powerful extensor mechanism which pulls out the pliable 'fan'. In Fig. 4 (p. 15) a wing of *Forficula auricularia* is shown in detail, and the lines along which the wing is folded are dotted. These show clearly that the wing folds up like a fan. It is remarkable how one particular point on every hard black vein is paler and weaker, 'pinched' as it were. This is the click point, indicated by the arrow in Fig. 50. This wing is provided with a folding device that is exceptionally elegant, complex and doubly reliable. Delicate though the wing may be, the air pressures during flight are never able to deform the wing membrane. If this ever happened the wing would snap back into the first stable position, automatically fold itself up, and cease to give any support to the insect. And this would be a disaster for an earwig, after it had eventually worked itself up to venturing into the air!

37. Migratory flights of locusts and butterflies

'Blow ye the trumpet in Zion, and sound an alarm in my holy mountain: let all the inhabitants of the land tremble: for the day of the Lord cometh, for it is nigh at hand; a day of darkness and of gloominess, a day of clouds and of thick darkness, as the morning spread upon the mountains: a great people and a strong; there hath not been ever the like, neither shall there be any more after it, even to the years of many generations. A fire devoureth before them; and behind them a flame burneth: the land is as the garden of Eden before them, and behind them a desolate wilderness; yea, and nothing shall escape them. The appearance of them is as the appearance of horses; and as horsemen, so shall they run. Like the noise of chariots on the tops of mountains shall they leap, like the noise of a flame of fire that devoureth the stubble, as a strong people set in battle array. Before their face the people shall be much pained; all faces shall gather blackness. They shall run like mighty men; they shall climb the wall like men of war; and they shall march every one on his ways, and they shall not break their ranks: neither shall one thrust another; they shall

Plate 27.   A bumble-bee (Bombus terrestris) *flying towards a bell of a fox-glove. The forward component of the body's velocity is almost nil, the abdomen hangs at an angle of 45° to the horizontal, and the wings beat in a nearly horizontal plane.*

Plate 28.   A bee flying into the entrance-hole of the hive, through which the photograph was taken. Note the masses of pollen carried on the corbiculae of the hind legs.

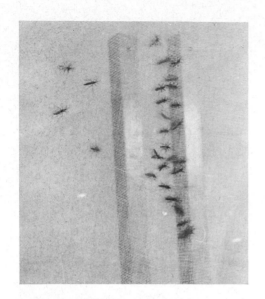

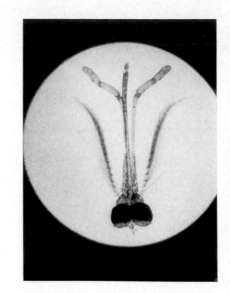

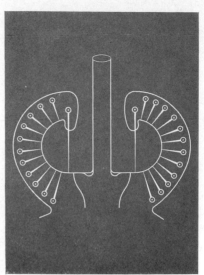

walk every one in his path: and when they fall upon the sword, they shall not be wounded. They shall run to and fro in the city; they shall run upon the wall, they shall climb up upon the houses; they shall enter in at the windows like a thief. The earth shall quake before them; the heavens shall tremble; the sun and the moon shall be dark, and the stars shall withdraw their shining.'

These words of the prophet Joel have a truly apocalyptic ring about them, and they inspired the illustrator of at least one medieval book to depict hair-raising visions (Plate 25). What he drew are clearly efforts of imagination, and were not meant to represent the knights of a hostile army, slaying and burning: in fact they are meant to be locusts!

Did not Joel exaggerate a little? Anyone who has ever seen for himself what frightful havoc these swarms create, and how impossible it is to fight against them, or to turn them aside from their destructive path, will think this description not at all exaggerated, but exactly fitting the facts.

A few years ago such a swarm of locusts descended upon the Plain of Seldon in northern Nigeria. Barely four hours later there was no more vegetation to be seen, but an area 25 by 5 kilometres was buried under an inch-thick layer of excrement. A big swarm may be enough to cover the entire area of Nigeria. They appear with a deafening roar, and blot out the sun. Eye-witnesses have compared the noise with the roaring of a mighty waterfall. The weight of locusts breaks off branches of trees, and the noise of their chewing sounds like the crackling of a fire. The destruction is complete, leaving the land looking as if it had been burnt. Stench and famine, poisoned wells and epidemics are the outcome.

Locusts of several species form swarms. The best known are the desert locust, *Schistocerca gregaria,* and the migratory locust, *Locusta migratoria,* while the Moroccan locust, *Dociostaurus maroccanus,* and several others swarm in smaller numbers. Locusts migrate in swarms only when they are in the migratory phase of their respective species, and under different conditions of crowding together or feeding, they may continue to live as solitary individuals. It is still not entirely understood why the nymphs (hoppers) start off by migrating long distances on foot, and then after their fifth and final moult assemble into huge swarms, and early one morning off they go. As the ground warms up after sunrise they begin to work themselves up to flight. First, individual locusts take wing, spiral round, and alight again. The movement is infectious and more and more take flight, until suddenly, with a roar, the entire swarm rises into the wing. If the wind blows steadily without convection, as it often does after rain or in thick weather, the swarm travels at an altitude of only 5 to 10 metres, and a dense swarm may contain anything from 1 to 10 locusts per cubic metre.

A typical example of the numbers present is that of a swarm, flying in a light breeze at a speed of 4 metres per second, and in which 2.5 million locusts pass over one hectare of ground each second (Fig. 51). This means that over an area equal to that of an average-sized living room there are about ten thousand locusts at any one moment. The speed of their flight over the ground is 15 kilometres per hour.

Swarms of locusts can also circle round and soar in thermals, sometimes rising several thousand metres without a single stroke of the wings, until the swarm looks like a cumulus cloud. From this point of vantage the insects can then glide a great many kilometres across country without any muscular movement, and thus their migrations can be accomplished with the minimum of effort. By the same process the swarm becomes spread out, so that widely dispersed swarms may have no more than one locust in ten to ten thousand cubic metres, or one locust in the space of an average gymnasium. Even so this may still mean a hundred thousand locusts per hectare of ground! Swarms always move downwind, and they tend to thicken up in the middle since the locusts on the outside turn and fly towards the mass.

Under certain circumstances a locust population that has been living for several generations in a locality may suddenly swarm and move out. For instance, a population of migratory locusts lived a static existence beside the river Niger to the west of Timbuktu, until in 1928 a series of massive increases of population began, and persisted for several years. By 1930 the locusts had already spread westwards to the Atlantic, by 1931 eastwards to Lake Chad, and by 1932 they had reached Egypt; in 1933 they drove through to South Africa, and by 1934 they swarmed over all Africa south of the Sahara.

Locusts are the best known of migratory insects, but butterflies also migrate. The available information about the migration of butterflies is gathered together in the little book by E. T. Nielsen entitled *Insects on the Move.* He includes some personal observations of the American butterfly *Ascia*

Plate 29. *The cockchafer* (Melolontha vulgaris) *spreads its wings* (bottom) *and takes off* (top). *The separate 'leaves' of the antennae are spread out. The wing cases (elytra) beat in rhythm with the wings, but their aerodynamic effort is nevertheless small because their strongly positive dihedral and backward tilt—although excellent for 'rowing' through the air—give them very little amplitude in which to beat and, even more important, they cannot make the complicated movements that are typical of membraneous wings. They can support only about 7 per cent of the weight of the insect at most. Pictures taken with Novoflex bellows and electronic flash.*

Plate 30. Top left: *male mosquitoes assembling on a tuning fork that is vibrating with the fundamental frequency of the wings of the female mosquito.* Top centre: *photomicrograph of the head of a male mosquito, showing proboscis and antennae.* Top right: *simplified diagram of the essential parts of Johnston's organ, explained in detail in Fig. 57.* Bottom picture: *an opera singer sings a glissando. As soon as he hits the fundamental note of the female mosquito he gets his mouth full of male mosquitoes!*

monuste, and, according to these, this butterfly can reach a speed of about 12 kilometres per hour, and cover distances up to 130 kilometres without pause. Only adult butterflies take part in the migration, and a cross-section of the swarm may include 4000–15,000 individuals. Sometimes the swarm arranges itself into a cylinder only about 2 metres in cross-section, and moves over the landscape like a monstrous flying serpent. As soon as the individual butterflies reach the age of 30 hours, the urge to migrate fades gradually into an urge to lay eggs. The females then break off their flight, settle on certain particular plants and begin to oviposit.

The migration of the monarch butterfly, *Danaus plexippus* is also well known. The caterpillars of this butterfly live only on the milkweed, and those butterflies that emerge in August or later often aggregate into huge swarms. In contrast to *Ascia,* described above, the monarchs fly southwards over a broad front, but they concentrate themselves into a particular area of central California, especially around the small township of Pacific Grove, a hundred kilometres south of San Francisco. Their arrival each year is commemorated by a big festival. The school children are given a holiday, and visitors come from far afield to witness the event. The town council of Pacific Grove has erected a large monument on the seashore in gratitude to the monarch butterfly, the first monument to a butterfly that I have ever seen.

Another migrant is the bogong moth which lives in Australia. The caterpillars of this moth inhabit the grassy plains of New South Wales, in the extreme south of the continent. The moths emerge in spring, then move upwards into the mountains and assemble in great masses in crevices. They frequently come out and fly around in swarms, but they do not feed, subsisting at the expense of a huge fat body carried in the abdomen. The aborigines were aware of this, and roasted the moths in heaps, the abdomens being both tasty and nutritious. Those moths that manage to survive the festive season leave their refuges in the autumn, pair, and then return to the lowlands again. Apparently the summer refuges in the crevices are assembly points for a mass pairing, though some authorities see the whole phenomenon as a device to escape the intense heat of the summer on the plains. The best-known long-range migrants in our own part of the world—Northern and Western Europe—are some of the big hawk moths—the convolvulus, elephant, Death's head, lime, oleander and spurge hawk moths—while Noctuidae, Arctiidae, Geometridae and Pyralidae also occasionally migrate. The convolvulus hawk moth is an inhabitant of the Mediterranean sub-region, and is rare in England, yet during the last 75 years no fewer than 5000 of them have turned up there. Occasionally an individual hawk moth even manages to reach Iceland. The Death's head migrates from the Mediterranean in two waves, one in May/June and the other in

Fig. 51. This swarm of desert locusts (Schistocerca gregaria) *seen in Ethiopia in 1958, covered no less than 400 square miles. Photogram taken from a black-and-white original.*

September. Every year some part of the population migrates northwards as far as Sweden, Finland and Iceland. On the American continent there are no fewer than 250 species of migrant butterflies, 22 of which appear in considerable swarms.

We have already talked about where the energy of flight comes from. Fat is a light-weight, concentrated fuel, which enables insects to make long, sustained flights without the need to 'refuel'. In a certain cicada fat makes up 40 per cent of the total body weight, but after flying 300 kilometres this percentage has fallen to 10 per cent; the difference is the amount of fat burned up to power the flight muscles. It has been roughly estimated that the desert locust has enough power available to enable it to fly about 350 kilometres, so it is capable of flying across the Mediterranean. Similarly the American monarch butterfly is equipped to fly 1000 kilometres at a stretch, but it does not do so.

Birds, and the golden plover in particular, are even better performers and flights of 500–700 kilometres are evidently quite normal. Individual unbroken flights of 1120 kilometres have been reported. Relatively small species of bats have been subjected to experiments in which they were transported 320 kilometres from home and returned to their roost without stopping, developing speeds of up to 30 kilometres per hour.

Furthermore, hummingbirds can fly across the Gulf of Mexico, which is about 800 kilometres across, and large hummingbirds can attain speeds of 80 kilometres per hour. Assuming that the necessary energy is obtained by burning fat, such a velocity would call for the expenditure of about 1.17 kilocalories per hour. If the bird is weighed before and after flight it is found to have lost in the region of 1.3 grams weight, which is accounted for by loss of fat representing 12 kilocalories of chemical energy. A hummingbird could fly at this rate for 10 hours ($= 12/1.17$) and could cover 800 kilometres at its average speed of about 80 kilometres per hour. The crossing of the Gulf of Mexico is thus within its cruising range. Many scientists have doubted the possibility of this, and have suggested instead that the hummingbirds really would turn south immediately and fly across the Straits of Florida towards Cuba, a distance of about 250 kilometres, and subsequently returning the same distance to Florida. Yet this feat would be impressive enough. Even the ruby-throated hummingbird is one of those that fly out over the Gulf.

All birds, and all the insects that have so far been studied, burn up fat. Carbohydrates are a much inferior source of energy as we have already seen in the chapter about flight organs. Honeybees and malaria mosquitoes can fly at most a distance of 50 kilometres, and horseflies and blackflies about 100 kilometres, because nectar as a foodstuff is heavier than fat, less nutritious, and can only be obtained in small amounts. The desert locust consumes only 0.8 per cent of its body weight per hour as fuel, *Drosophila* consumes 10 per cent, the honeybee 30 per cent and blowflies 35 per cent. Light aircraft and helicopters, for comparison, consume up to 2–5 per cent of their body total weight per hour, jet airliners 12 per cent and supersonic fighters up to 36 per cent. Both natural and man-made flying-machines that have a long cruising range are designed to be economic in fuel consumption, whereas those which are able to refuel frequently are able to get by with a higher fuel consumption. For example, the bees, flitting from flower to flower, top-up frequently, and so make use of nectar, a foodstuff less rich in energy than fat.

When it comes to supersonic fighters the rate of fuel consumption is not the limiting factor because they stay in the air only a short time. The essential thing is that they should be fast, and highly manoeuvrable, and to achieve this an adverse energy-balance is acceptable. Fighting machines are not expected to be economical!

Finally we have a few interesting observations on the cost of power. Suppose that we compare the relative amounts of fuel needed to transport one kilogram at the normal speed of each particular vehicle. A big propeller airliner might need 1 calorie; a jet airliner 2; a small light aircraft 2–3; a helicopter 4; a supersonic fighter up to 5. A desert locust would also need up to 5 calories, a fruit fly *(Drosophila)* 7; and finally blowflies and honeybees up to 30 calories. Clearly, therefore, insects are not efficient load carriers. But then they were not designed to meet human specifications.

38. The migrant habits of our native insects

Locust swarms come our way only exceptionally, and the routes followed by hawk moths from the Mediterranean are well known. Much less is known about the migratory habits of our native insects, aphids, bugs, midges, ladybirds, dragonflies and what have you. I am calling these aggregations of insects 'native' although their migratory habits are still unknown to most people, in spite of the fact that these creatures often appear in great masses in widely separated localities. Aphids are a typical example. They migrate quite regularly, every year, often in hundreds of thousands. But have you ever seen a swarm of aphids?

Each species of aphid shows a preference for feeding on one particular host plant, and a typical life cycle would be as follows: from the eggs hatch the young aphid nymphs, which are fully grown up after four moults. They are then wingless females. In due course each of them, without intercourse with a male, produces three or four offspring parthenogenetically, and these in turn repeat the process as soon as they are mature enough. As a result dense colonies of wingless female aphids grow up. Later in the year winged forms arise, still by parthenogenesis; these are the migrating generations which fly off in search of another species of plant, called the alternative host. There they produce young which may be either winged or wingless, but which are still all females. Towards autumn only winged generations appear, and these fly back to their original host. Males now make their appearance, and pairing takes place; the females lay their eggs, and the parents die. The eggs overwinter, and next year the whole cycle begins again.

Aphids often fly away in search of an alternative host, and in the process swarm in immense masses. Sometimes the air is thick with millions of individuals. They are literally 'gone with the wind', since they are too weak to control the direction of their flight themselves. They are most often in movement from mid-morning to mid-afternoon. All of them land again before sundown, and no aphids fly at night. They fly horizontally, low above the ground, and often make intermediate landings to sample the vegetation. In their brief halts they always deposit a few young nymphs, and refresh themselves by sucking a little sap from the plant. At the end of a migratory flight they prefer to feed on young, sappy vegetation, and the shorter the flight has been, the more critically they choose where they will feed. After a long, exhausting flight they will accept vegetation that is less succulent, sometimes even species of plant upon which they cannot usually get enough nourishment.

Ladybirds are occasionally found in large numbers. In good weather in the autumn ladybirds are inclined to form swarms and move off into special winter quarters, often near the tops of mountains. An aggregation of this type was once seen on Mount Etna, where virtually every lump of lava had its little beetle sitting on it. Probably ladybirds choose high altitudes because it remains cold for longer up there, whereas down in the valleys there are often short intermissions of warmer weather. These would arouse the laydbirds from their winter sleep, and the next cold day would freeze them to death. Upon the mountain tops there is little risk of this.

Freshly emerged mosquitoes are ready for their first flight four or five hours later. They move about only in the dark, and for this reason they wait until dusk before showing any activity. Shortly before sundown there is a general stirring. First the mosquitoes make short flights from twig to twig, but always moving upwards towards the treetops. About a quarter of an hour later most of them are up at the top. A few dozen of them rise up from the top of the tree like a puff of smoke from a chimney. Fifteen to thirty seconds later comes the next wave, and so it goes on intermittently. The mosquitoes mostly fly only 10–20 kilometres at the outside. These distances have been recorded using radioactive isotopes as markers. In Florida a hurricane flooded the fields, and in the warm still water myriads of mosquito larvae quickly appeared. Naturally all these larvae reached maturity and emerged as adult mosquitoes at about the same time. They rolled like a carpet across the countryside for a distance of about two dozen kilometres, always keeping about one metre above the treetops.

Dragonflies, too, may succumb to wanderlust. In 1852 a massive movement of large dragonflies was observed near Königsberg, 15–20 metres broad, and 2 metres high. They moved steadily at the speed of a slowly trotting horse, directly across fields and meadows. The astonished observer saw dragonflies passing him at the rate of 20 million per hour from 9 a.m. till dusk, without any falling off in numbers. The mass of wings glistened in rainbow colours, indicating that the dragonflies were freshly emerged. This migration away from the place of emergence of the dragonflies led towards a very small and still sheet of water to the north-east of the town. For this hundreds of millions of dragonfly larvae had spent their lives, had climbed to the tops of reed stems, and had simultaneously released their millions of glittering adults. Low over the ground the migrants had moved with military precision, not deviating a hair's breadth from their path.

Migrations of dragonflies take place only under certain particular conditions: either when enormous numbers emerge all at the same time, or else in smaller groups as they grow older, right up to the time when they are no longer sexually potent. But swarms on the scale of that at Königsberg come along only about once in fifty years. These compulsive migrations are fascinating, and it is by no means fully understood what causes the insects to act simultaneously.

Bees often fly as much as a kilometre from their hive when they are looking for a crop of flowers that is rich in nectar and pollen, and then the heavily laden bees have to find their way back to the hive. From a distance of a kilometre, or even a few hundred metres, the hive is no longer visible. How, in fact, do the bees manage to find it again?

Obviously it is advisable for the bees to get their loads back to the hive as quickly as possible, and by the shortest route, preferably a straight line (a 'bee-line'). It is a waste of energy to have to search for the hive, and a bee's supply of fuel is limited, particularly since the carbohydrate nectar is inferior to fat as a source of energy. The more uncertain a bee's navigation is, the more time it spends in searching for the hive, and the more of its load is burned up on the return journey, and hence the lower is its payload. Moreover, if the bee does not know its way it cannot tell the other bees where to go to find the rich source of nectar and pollen. It is now known that bees can communicate with each other. The returning forager performs a dance in the hive, which tells the other bees where to look for the source of food (Fig. 52). These bees then pour out of the hive, and unfailingly locate the food supply, provided that they have been given the correct information.

All this comes back eventually to the question: exactly how does a bee orientate herself during her flights between the hive and a source of food? Considerable distances are involved, which need to be looked at from the bee's standpoint, and not our own. A kilometre is not very far for us, only 550–600 lengths for a grown man. For a bee one kilometre is 66,000 body lengths!

If a young bee, one which has never been out of the hive before, is taken off the honeycomb and put down somewhere outside, only a few dozen metres away from the hive, she cannot find her way back. It is quite a different matter if she leaves the hive voluntarily when she is ten days old. Her first short flight is solely for the purpose of learning her way about, and she does not collect any nectar or pollen. She is fully occupied with leaving the shelter of the hive and memorising landmarks round about it, and then she can always find her way back to the hive again. A few such flights are enough for her to learn all the features within a kilometre or so of the hive, and to have committed them to memory.

If several hives stand close together the bee orientates itself by the position, colour and shape of its own hive, as well as

Fig. 52.   The tail-wagging dance of the honeybee. The dancer bee goes through the routine many times, each time wagging her abdomen quickly to and fro as she traverses the central straight section of the pattern. At the same time she makes a noise, which is produced by the flight muscles of the thorax. She then returns in a semicircular path to the starting-point and she may repeat this dance up to 200 times. Foraging bees follow after her. Directional information is conveyed by the oreintation of the straight part of the pattern in relation to the vertical (cf. Fig. 54). Distance information is conveyed (according to Karl von Frisch) merely by the tempo of the dance (up to 100 metres distance is conveyed by a rate of 40 dance figures per minute; 500 metres by 24; 5000 by a mere 8). There are still other parameters, however, that are correlated with distance. Barely twenty years after the ingenious researches of von Frisch, H. Esch showed by precise experiment that the number of tail-wags per dance (and thence, since the frequency of tail-wagging remained constant at 15 Hz, also the time taken by each tail-wag) was correlated with increasing distance of the food. The low-frequency tail-wagging was accompanied by a high-frequency sound vibration (250 Hz), at which frequency the thoracic muscles could vibrate the wings. This activity diminished exactly in step with the frequency of the tail-wagging phase, and hence must be correlated in the same way with the distance of the food. Undoubtedly these high-frequency sound signals play an equally important part to the tail-wagging in conveying distance information, if indeed they do not themselves constitute the actual information. Observations made on primitive relations of the honeybee, the stingless bees, show clearly that it is justifiable to speak of the 'evolution of bee-language' (Fig. 54). Pointing in the same direction are some incidental observations on individual bees which also dance but for some reason or another cannot produce the high-frequency sound. These individuals are not able to direct bees to a source of food.

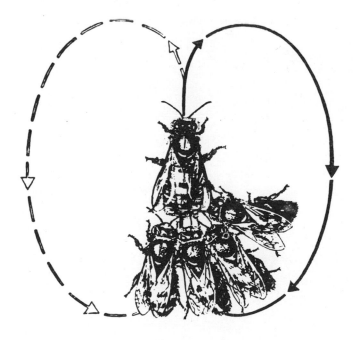

by its typical nest smell. This is 'short-range navigation', and 'long-range navigation' requires other landmarks, but these alone are evidently not enough, because bees have evolved a sophisticated system of co-ordinates for indicating a particular locality in which food is abundant. This system is more than just a means of orientation, and is still more important in providing a means of mutal understanding between one bee and another. The returning forager cannot just say to her neighbour on the comb: 'I have discovered a mass of nectar. If you want to share it, fly to the third willow tree on the left, then 60 metres along the brook, behind the beech clump, then turn right down to the bridge, over the field of yellow rape, and then diagonally across to the little pond. Just to the left of that you will find the flowers that are in bloom.' This might be a short way of passing on the information, but in the example given there are already at least a dozen separate instructions.

There are systems known by which this information could be given in only two items. One such system is the use of polar co-ordinates, which is all very well for airliners which are able to fly in a direct line to their destination. Make a spot on a piece of transparent paper, and draw four lines from it at right angles to each other, as in Fig. 53. Label one line 'N' for north, and place the sheet of transparent paper over a street plan of your home district. Locate the centre M over your own house, and let the line N lie exactly north. Find your favourite pub, and mark it with a spot K. Suppose the relative positions are as in this sketch (Fig. 53). To express how the pub lies from your house requires two co-ordinates: the length of the line M–K in metres, and the angle α in degrees. In the sketch M–K is just 300 metres, and the angle α is 45°.

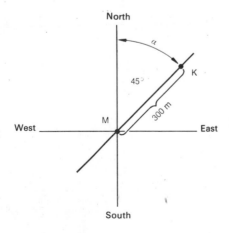

Fig. 53. *Diagram of a polar co-ordinate system. M = starting point; K = destination; N = direction of true north; α = directional angle (see text).*

These are the two components of the polar co-ordinate system.

We have just measured the distance in metres and the angle as a compass bearing in relation to true north. The bee probably measures distance by the amount of fuel (carbohydrate) that it has burned up, and determines the angle by 'sun-compass'. It lays off, as it were, a line from the willow tree to the hive on the one hand, and a line towards the sun on the other (it does not matter how high the sun is in the sky), measures the included angle, and takes note of it (Fig. 54). On its return to the nest the bee can pass on both pieces of information to other bees in the hive by means of a 'tail-wagging dance'. The precise way in which this is done can be read in detail in any one of the elegant popular works of Karl von Fritsch or in the more recent papers of Harold Esch and other American authors. We are not concerned to that extent in the present work, as long as we understand that the approximate distance of the food supply is indicated by the frequency of the 'tail wagging', and the approximate direction by the direction of the middle axis of the dance pattern.

Astonishing as it may seem, the bees inside the hive understand the information exactly. They fly away from the hive in precisely the right direction, which in our example is north-east, steering by the sun, and exactly 300 metres away the bees come down to earth, automatically finding themselves at the feeding-place.

This is a biological masterpiece, making the most of a minimum of information, and is also a masterpiece of precision. Even a novice bee, interpreting the dance of a forager for the first time, is accurate within an angle of ±8° (ie an error of only 2.2 per cent of a complete circle) and within an error of, for example ±50 metres in 1000 metres, or 5 per cent error in judgement of distance. These errors include misunderstandings between the forager and the novice bee as well as inaccuracies in the orientation of the dance on the comb.

It seems to be a weakness of the system that the bee does not use the magnetic compass, but has to rely on taking a direction from the sun. Magnetic directions remain fixed, whereas the sun moves round all day. If a bee flies away from the hive at an angle of 45° from north as directed but steering by the sun, collects nectar and pollen for a time, and then returns on the reciprocal of 45°, it will not get back home to the hive, and will have lost its bearings. How can a bee compensate for the apparent movement of the sun during the foraging trip? Only by measuring the time elapsed and allowing for a steady movement of the sun of about 15° per hour. This information is apparently inborn in the bee, which reads off the time on its own internal clock.

To take a practical example. The bee flies at an angle of

45° to the left of the line from hive to sun, collects for 30 minutes, and then wants to fly back to the hive. At what angle to the sun must it now fly? In 30 minutes the sun has moved 7.5° towards the west, so the bee must make an angle with the sun (azimuth, in astronomical terms) of $45 - 7.5 = 37.5°$. If the bee does this it will find its way back to the hive. How does the bee know how long it has spent foraging at the feeding-site? Have bees pocket watches? In a certain sense they have, and carry them always on their person! We call such devices 'biological clocks', but we do not know either how a clock works, nor how the bee can read the time from it. Max Remer pertinently asks: 'Do cyclical biological processes constitute the hour and minute hands?'

The following explanation shows how amazingly accurate the bee's clock is. Bees sometimes dance for a very long time on the comb. In particular the so-called 'scout bees', which look out for a suitable place for a new nest before a swarm of bees leaves the old hive, may dance for as long as an hour, with pauses. During this period they do not come out into the daylight. The protracted dance serves the purpose of allowing every individual bee to have a chance to learn in what direction the swarm will fly when eventually it emerges from the hive.

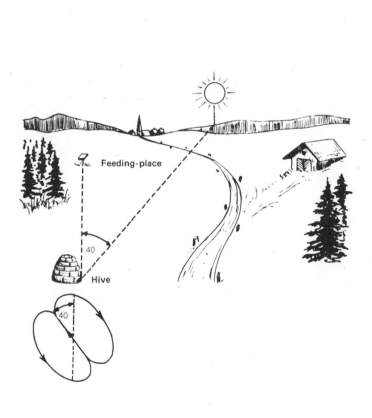

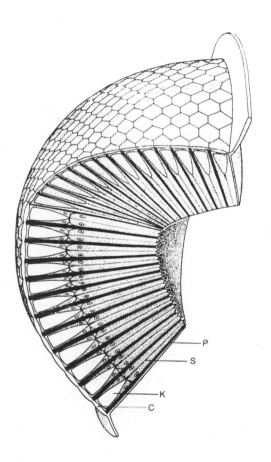

Fig. 54. *Sketch of direction-finding and the conveyance of directional information by the honeybee* Apis mellifera. *If the bee finds that the direction of the feeding place makes a horizontal angle of 40° to the left of the sun (the sun's height in the sky is irrelevant), then subsequently the bee dances on the vertical comb in such a manner that the straight centre track of the dance pattern (Fig. 52) is at 40° to the left of the vertical. The bee has converted the angle into different terms of reference, a gravitational system, because the sun is not visible inside the hive. Primitive bees of the species* Apis florae *live in the open and only dance on horizontal surfaces and directly in relation to the sun; they do not transpose their co-ordinates. The honeybee itself occasionally dances on the alighting board of the hive, and can easily be made to do so inside the hive if the comb is turned to a horizontal position. Still more primitive genera of bees do not dance at all but induce other bees to fly after them directly towards the food, or else mark the route every few metres by scent-spots on the ground (*Trigona postica*). It is reasonable to assume that the 'evolution of bee language' has taken place from such primitive ways of conveying directional information up to the complex transposition system of the honeybee, the most highly organised community of any insect.*

Fig. 55. *Diagram of the hemispherical, faceted, apposition eye of the insect. One sector is shown in section. The most important parts of the many wedge-shaped ocular units, or ommatidia, are the externally hexagonal sclerotised facets, or cornea (C), the crystalline cone (K) which directs the light, and a group of sensory cells which in cross-section looks like a section of an orange, and their light-sensitive inner cores, the retinulae (S) which touch each other centrally. Often these receptor organs are fused with a central rod, the so-called rhabdome. Dark pigment cells (P) isolate the ommatidia optically from each other, and by this means make a sharp image possible. If the above interpretation is correct, each ommatidium must produce only a single point image of what is in front of it, in which it records the energy content (brightness) the wavelength (colour) and—if polarised light is present—the direction of rotation of the light. Even ultraviolet light and X-rays are detected. (According to the latest research, six cells from the adjacent ommatidia, though morphologically less important, are directed towards the same point on the image, and so the seven constitute a single unit, a 'physiological ommatidium'.)*

However, during the hour-long dance the sun has moved round the sky, at a rate of 15° per hour. Minute by minute the dancing bee alters the angle of its dance, compensating for the slow rotation of the earth, and keeping its indicated direction correct by the sun. She does this without any glimpse of the sun itself, with the precision of clockwork, in fact under the control of her own internal clock.

Thus the bee has all four components, the four parameters necessary to measure direction, transform it into polar coordinates, and communicate it to other bees: 1 a measure of distance by fuel consumption; 2 a compass, the sun; 3 a timing mechanism, her own internal clock from which she can correct item 2, the sun angle; and 4 a language, by means of which she can communicate items 1 and 2 to other bees.

Which of these abilities is the most amazing? The incredibly simple and effective conversion into polar coordinates? or the bees themselves that have evolved such a technique? or the scientists who have unravelled this process?

It must be added that the bees can still determine direction from the sun even if the sun is not visible. This is only an apparent paradox. The sun creates a characteristic pattern of polarised light all over the sky. We ourselves cannot see polarised light, but the bees can, with their facetted compound eyes (Fig. 55). It is enough for them if a gap in the clouds allows them to see even a small area of sky. They can then make a reliable deduction of the sun's position in the sky from the fragment of polarised pattern that is visible to them. We can demonstrate something similar by cutting a small hole in a newspaper and placing it over the face of the kitchen clock so that only one figure is visible—for example the figure 2. From the location of the '2' and our secret knowledge of the dimensions of the clock face we can work out where the centre of the hands comes under the paper.

There are many other problems about how flying animals orientate themselves, though some of these problems have already been solved. Migrating birds, for instance, may use sun, moon and stars as navigational aids. Possibly too, the earth's magnetic field may play some part. Whatever the answer may be, such questions are among the most fascinating as well as the most important aspects of behavioural research.

40. How is flight velocity regulated?

The problem of orientation is closely coupled with another problem: how is flight velocity regulated? There is a great deal to be said for the idea that bees estimate the distance they have flown from the hive on the basis of fuel consumption, but since demand for fuel rises and falls with the speed of movement there are two quantities on which fuel consumption depends: the flight distance and the flight velocity. If you want to estimate distance by means of fuel consumption, you must of course be able to relate it to a particular exact velocity.

A car driver who says 'I get 30 miles per gallon' is referring, let us say, to a constant average velocity of 50 miles an hour. As long as this velocity remains constant, he can calculate any distance directly from his fuel gauge. He knows that when he has used up a gallon he is 30 miles from his starting point, when two gallons are gone he has driven 60 miles, and so on. On the other hand, if he has driven the whole distance with the pedal on the floor, averaging 70 m.p.h., then he would say, perhaps, 'I'm now getting 20 miles per gallon'. And the calculation of distance covered on the basis of fuel consumption would have to change accordingly. After using one gallon he would now be only 20 miles from the starting position and after two, only 40 miles.

If the driver wants to hold his velocity constant, he depends on a measuring instrument, the speedometer. He reads it

Plate 31. Above: *a blowfly of the genus* Calliphora *is in the act of landing on a piece of meat. She rushes up to it at an incredible speed, and on the upstroke the wings bend sharply downwards. They have a preformed line of weakness that makes this possible. A 'heavy landing' such as this is not really typical of the fly, but it shows particularly well how extremely manoeuvrable and quick in its reactions is such a 'clumsy' fly as the blowfly.*

frequently and repeatedly adjusts the accelerator with his leg muscles so that regardless of whether he is driving uphill or downhill, with or against the wind, the speedometer pointer is always on, say, 50. Of course, one could also easily build an electrical servo mechanism which would automatically hold the velocity reading of the speedometer constant by giving more fuel whenever the pointer begins to fall and less when it moves too far up. This would greatly relieve the brain of the driver. He would no longer have to be constantly paying attention, making comparisons, and giving the proper orders for adjustment. And it is just such a completely automatic measuring apparatus as this which the bees and flies use: their antennae.

Let's make a quick thought-experiment, by inventing, as in Fig. 56, a constant-velocity-regulator for a car. We want to make everything as simple as possible. To measure the velocity, we take a paper plate, and suspend it from above. The plate is pushed backwards by the resistance of the air as the car moves forward, and it moves further back the faster the car goes. If we wish, we may set up a scale under it to show the displacement in terms of miles per hour. This velocity meter is coupled by a rod to the movable contact of a sliding-wire potentiometer, and acts against a restoring spring. The potentiometer is connected in such a way that it decreases the voltage of a battery as it is slid to the right. There is an electromagnet in the circuit which pulls a string hooked onto the accelerator pedal. And that's all we need: when the velocity increases, the air stream produced also increases and pushes the plate to the right, increasing the resistance. As a consequence of this, the voltage at the magnet becomes smaller, so that it doesn't pull so hard, and the spring attached to the accelerator moves it further back. The result: the car goes slower, until it has once again reached its original velocity. If the velocity should continue to decrease, for example if the road runs uphill, then this regulatory mechanism will provide more voltage to the magnet, which will pull more strongly on the pedal, and the car will go faster.

The technician calls such a system, in which several elements are connected in a circuit, a 'control (or regulatory or servo) system'. The value at which the velocity, in our example, should be maintained is called the 'set point' (or 'reference input'). Our measuring instrument, the paper plate, is in general called a 'monitor'—in insects the antennae serve as monitors (Fig. 57). The accelerator, which determines the force exerted by the motor, is the 'control element'. In insects the control element is the mechanism which sets the amplitude of the wing movement. Since the more the plate is displaced, the less fuel is eventually given, that is to say, the influence exerted by the plate is negative, this kind of coupling between monitor and control element is called 'negative feedback'.

In nature it is the insect antenna that is displaced. The negative of its signal is sent to the mechanism regulating wing-beat amplitude, by way of a sense organ in the antenna, then via the nerves to computing centres in the central nervous system, to nerves once again and finally to the flight muscles. If the insect antenna is bent more strongly, the wing-beat amplitude is made smaller as a result, and thus the undesirable increase in velocity is reduced until the velocity reaches its set value, for example 14 m.p.h. We can see that, from the point of view of regulator technology, the two systems are entirely identical. The result in either case is that any perturbation of the travelling speed is automatically compensated. Amount of load, smoothness of the road, inclines and declines have all become irrelevant. Now, for the first time, one can really use the fuel gauge as a distance meter. The bee does this all the time.

Still, this scheme has one little flaw, both in nature and in technology. If either the driver or the insect should want to change the set velocity, when for some reason they wish to move more slowly or quickly, they cannot. The completely automatic control loops hold the reference velocity firmly at 50 or 14 m.p.h., respectively. Of course, one could switch off the control system and take over the controls oneself. But there is another, more elegant, possibility, whereby the control system is left quite intact but yet forced to work towards whatever other velocity may be desired. The bees have been using this device for millions of years; the technologists, only for decades. It is called 'set-point selection', and it works very well.

The monitor in the control circuit is tuned with respect to a particular set point, and it is this value alone which the circuit is capable of matching. But if one changes the monitor so as to give it a new set point, for example 70 rather than 50 m.p.h., then the circuit accepts this new value and regulates itself automatically to match it. That is the whole trick; and with it one has, so to speak, outsmarted the control circuit. It performs its extremely practical service as well as before, and we can now dictate what it does.

Suppose that a small lead weight were to be glued to the bottom edge of the paper plate. What happens then? We

137

have given the system a new set velocity. This is easy to understand. The disc has been made a little heavier and as a result, in spite of the air stream produced by 50 m.p.h. movement, it falls back somewhat towards the vertical position. This raises the voltage at the solenoid, the accelerator is more strongly depressed, and the car goes faster until the new set point has been reached. When that happens, the circuit zeros in on this new velocity and maintains it.

Velocity alteration by changing the set point—there we have it! The bee does it in just this way. It has particular muscles which can change the position of the antennae slightly with respect to the air stream. Then the air resistance during flight can no longer bend them by the same amount as before but rather, for example, a little less. This is just the way the extra weight on the plate worked. The result is also

the same: a new set point for the flight velocity is achieved. As a result there is a change in the output to the wings which produces an altered wing-beat amplitude. The motive forces change and the bee zeros in on a new velocity. In this way it 'kills two birds with one stone': it holds the set velocity constant, but it can select any set point it likes. Bees and flies have two antennae, one right and one left. It has been shown that each of these antennae regulates only the amplitude of the wing on its own side of the body. This has possibilities for flying in a curved course. If one antenna is cut off, the insect always flies around in circles. It is only when both antennae are operating together that they permit the bee to fly in a straight line and compensate for gusts of wind which might push it a little away from this line. This is critically important for orientation of the flight between hive and food

Fig. 56. Schematic diagram of a mechanism for regulating the speed of a car so as to hold it constant. Detailed discussion in the text

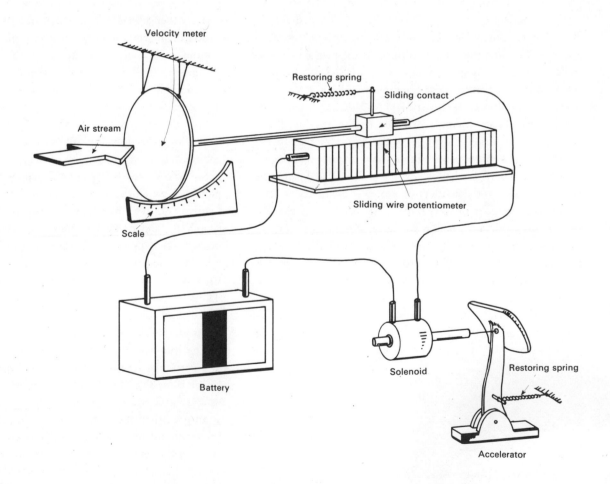

source, which the bee should make as straight as possible.

Let us stop at these fundamental observations: in practice the whole thing is considerably more complicated. But there is one more point to be made: the bee does have a second servo system to control its velocity, involving the two large compound eyes (Fig. 55). The essential purpose of this system is to hold constant the velocity over the ground; the antennae, on the other hand, regulate the velocity through the air. When there is no wind these two velocities are equal: if the bee flies through the air at 5 metres per second it naturally proceeds in one second over 5 metres of ground. But the situation is different when there is a head, tail or side wind. Suppose the busy gatherer is flying at 5 metres per second through the air, but there is a tail wind with the same velocity. Then even though it moves with respect to the air at 5 metres per second, the velocity over the ground is twice as great. What would happen if the bee had only its antennal control circuit?

One simple example: the bee is getting ready for a flight to a gathering place with which it is already familiar. For this known distance it has calculated that it needs a particular amount of fuel (sugar). So it says to itself, 'I'll start off now in such-and-such a direction and fly until I've used up a cup of sugar. Then I'll land and there will be my flowers. That has worked three times before and must work again now!' The bee takes bearings by the sun, selects the proper set velocity of 8.2 metres per second, and flies away. Suppose that there had been no wind when it made the first three flights, but now there is an unexpected tail wind of 5 metres per second. The antennae regulate the air speed to be just 8.2 metres per second, as they should do. They can't tell that in one second the bee is being pushed an extra 5 metres over the ground. They can only measure the air resistance, and that is always proportional to the difference in velocity between the bee and the surrounding air, not between bee and ground. So the bee flies happily on, using the same amount of fuel as it had done in still air, but now in one second it progresses 13.2 metres rather than, as it must think from the antennal signals, 8.2 metres. When it has used up its cup of sugar, it lands. But where does it find itself? It has long since flown past the target. Assuming that the distance of the flowers from the hive was 820 metres, it will have landed at a distance of 1320 metres—a complete failure.

It could have avoided this embarrassment if it had paid more attention to its ground speed rather than air speed. And that is in fact what it does. For the physical reasons we have discussed, an air-stream monitor such as the antenna (or paper plate) cannot measure the ground velocity. But an optical measuring device which fixes points on the ground and computes how quickly they move backwards, can do this. The highly complicated compound eyes are admirably suited to this task. They can recognise the dangerous drifts produced by head, tail or side winds and compensate for the error in the signals of the antennal control circuit.

We could go on to discuss the ways this is achieved in circuit technology, but we don't want to get involved with electronic details. The important message here, in summary, is that bees possess two kinds of velocity meters, one which monitors the air stream and an optical one. When the air is still the first is sufficient, but when there is a wind the second must provide a correction. It is always advisable to have a second string.

Reconnaissance planes have used a similar double-insurance arrangement since the Second World War. Normally the magnetic compass is adequate, but when there are magnetic disturbances the gyroscope must provide corrections to the course.

41. Sense organs are the insect's flight control instruments

We have already learned about the most important control instrument, how bees and flies regulate their air speed through the sense organs of the antennae. If we think how complicated is the instrument panel of an airliner, we should normally expect that an entomological flying-machine would also have a multiplicity of measuring and recording instruments, controls and dials. Oddly enough there don't seem to be as many of these.

The rhythmical excitation which reaches the flight muscles in the form of nerve impulses does not arise from signals received by the antennae, but arises in certain particular cell complexes of the central nervous system, known as ganglia. These ganglia have the peculiarity of being able to generate such periodic discharges entirely without outside stimulus. The sense organs situated externally on the insect's body have no power to influence the discharges, being able neither to initiate them nor to control their direction, nor to regulate them, within quite wide limits. The control of flight operates from inside outwards, and for this reason is called 'endogenous rhythm'. The role of external sense organs is restricted principally to giving information about the aerodynamic forces acting on the insect during flight, to regulating the speed of flight, and to controlling the insect's attitude in the air.

The locust carries on the frons, or upper part of its head, a number of hairs which respond to air flow. These sense organs inform the central nervous system after take-off that the flight has begun, and is being continued. They also measure the direction of air flow relative to the head: they react to side-winds such as may arise from sudden gusts or when the insect is turning, and the information they supply enables the central nervous system to adjust the direction of flight accordingly. The antennae of flies and bees work in a similar way, but more precisely and reliably. The honeybee, and many flies, have the finest of hairs standing between the hexagonal facets of the compound eyes, and these hairs are sensitive to air flow, and certainly play some part in directing flight. So in the eyes we have an optical and a mechanical sensory system combined.

Sense organs in the tarsi of the fore legs of flies signal to the central nervous system at the start of the flight, 'contact with the ground lost', and so contribute to the process of launching into flight. We have already spoken of the special part played by the compound eyes in regulating the speed of flight over the ground, and any necessary compensation for side winds. Bees are 'visual fliers' of this type, whereas dragonflies have a tuft of sensory hairs on the prothorax which checks that the body is moving straight through the air, and can influence the direction of the wing stroke through a kind of servo mechanism. This provides an effective means of controlling the direction of the power stroke of the wings. Optical, acoustic and olfactory stimuli have an indirect, but considerable, influence on the flight of the insect. For example, male mosquitoes home on to the local sound source created by the flight tone of the female, as we shall hear more about in the next chapter. Locusts have sense organs on the wings which transmit either a single impulse or a short salvo to the central nervous system whenever the wings are almost vertical (cf. Fig. 39, p. 96). Other sense organs report when the wings reach their lowest position. If these so-called strain recorders are destroyed the locust can fly almost as

well, but more slowly, and with the wing beat less steady; the flight is not as powerful, and soon dwindles away to rest. Thus the strain recorders do not control the wing beat directly, but stimulate the central nervous system into an orderly control of flight activity. They act as stimulators, and monitors, and as control organs for the position of the wings.

A whole series of similar sensillae exists at the bases of the wings, and these respond to mechanical stimuli. Many things point to the likelihood that they act collectively as monitors of lift and drag, and of the twisting of the wing. All these components extend along, and penetrate into, the thick veins at the wing base in varying ways. The sensoria are correspondingly distributed. Such an array might work like a man-made strain-meter, and if several were orientated in different directions along an object, and reacted against each other electronically, any given deformation of the object could be resolved into its components, and thereby measured.

Two-winged flies have, in addition, a remarkable mechanism called the balancers, or halteres. These look like little drumsticks, and beat up and down with the same frequency as the wings. It used to be thought that these fast-vibrating structures acted as flywheels, stabilising the fly mechanically. If they are cut off, the fly flies unsteadily, and tumbles over in the air. Calculation shows, however, that even though the vibrating haltere is relatively big compared with the fly's body, it is still far too small to be able to stabilise it mechanically. Nowadays opinion is inclined to the view that the halteres constitute an organ for control of attitude in the air. It works like the automatic pilot of an aeroplane, which is a gyro compass linked to a servo mechanism, and controls the aeroplane's attitude along all three axes of rotation. If the fly deviates suddenly from its path, the central nervous system is probably informed at once in which direction the deviation has occurred, and with what angular velocity. At first glance this might seem unimportant, because the compound eyes would eventually tell the fly in what direction it had turned, but conditions are different in twilight or even in absolute darkness. The halteres give the fly a 'dynamic stability', so that it can fly under control even in darkness though it does not do so from choice. Bees have no halteres, and cannot manage a reasonable stability in the dark.

That is all the measuring instruments that are available to monitor the flight of insects, as far as we know:

Sensors register 'contact with the ground lost', 'wings raised', and 'wings lowered'.

Detectors record 'flight continues normally in a straight path', 'side wind right', 'side wind left', and 'body is at such-and-such an angle to gravity'.

Measuring equipment for registering and regulating airspeed and ground speed.

Possibly *stress gauges* for indirect measurement of the components of wind force, and of the geometrical twisting of each wing independently.

A complicated *autopilot* in flies, for dynamic stability during flight.

There is much here that needs further investigation, but this is also one of the most difficult fields of study for experimental work. Up until now the reliable information available is still all too scanty, and this is a matter of regret as much to the man who likes to build equipment for making experimental measurements as to the biologist himself.

42. By their song ye shall know them

Male mosquitoes dancing up and down in huge swarms, and in obvious harmony, are not dancing for pleasure: they just all happen to be doing the same thing at the same time. They are an assembly of individuals, each with all his senses alert, on the lookout for a female and ready to go after her without delay, and to mate with her before his neighbour gets hold of her first. The expression 'on the lookout' is not quite accurate, since the males do not locate the females by sight, but by hearing. They recognise the fair sex by her flight tone. The males themselves, however, also emit a flight tone, and each of their hundreds of neighbours is doing the same. How can they possibly pick out from all this din the sound of a single female as soon as she enters the swarm? The male's own wings alone are creating much more noise than is emitted by a female some distance away, because the flying machinery is located only a few millimetres away from the male's own hearing organs. None the less the male does find the female, and with a quite amazing certainty. Anyone can see this for himself, if he can bear to station himself in the midst of a swarm for a quarter of an hour! Wesenberg-Lund described it thus:

'One fine evening towards the end of September, numerous small swarms of the gnat, *Culex pipiens,* consisting entirely of males, were hovering in the shade of a lime tree in my garden. Around six o'clock the insects totalled only a few hundred, but between seven and eight o'clock they had increased to several thousands. The swarms hovered in the same spot for some time, and were usually columnar in shape, about 2–3 metres high, and about 1 metre thick, and they hovered about 5–6 metres above the ground. The individual gnats flew incessantly up and down, while the entire swarm swayed to and fro in the wildest excitement, attracting attention by its rapidly accelerating tempo. In the course of about three hours I saw the number of swarming gnats increase five-fold from individuals that flew straight into a swarm from outside. It seemed as if they were drawn together by some magic power, presumably by sound, but it was a sound that I could not hear. As soon as a female entered the swarm she became a focus of attention, and immediately the female and one male would fly out of the swarm together, descend together to the grass, and there pair, after which they both flew away separately.'

The problem of how they recognise each other was solved independently by several research workers. Horst Tischner wrote of this phenomenon:

'If a female flies past the swarm at a distance of 1–2 metres, several male gnats will leave the swarm and fly towards her. The males have detected with their antennae the flight tone of the female, though it differs only relatively little from their own, being about 300 Hz, whereas the males beat their wings about 500 times per second, and so have a note of 500 Hz. The antennae of male gnats have a natural period of vibration of about 300 Hz, so that they resonate to the sound of the female transmitter, and track her movements through the action of a relatively huge electro-mechanical receiver, called Johnston's organ.'

Fig. 57 explains the construction and working of this sensory organ. It is not difficult to produce a sound pitched at the resonant frequency of the antenna, either from a loud-speaker or by singing, so that the males will be deceived into thinking that a female is nearby. A few seconds later several male gnats will fly into the singer's mouth! Experiments in attracting males by means of a tuning-fork leads to similar results.

The secret of this amazing orientation mechanism is 'resonant frequency'. The antennae of the males are tuned to the 300 Hz emitted by the females, but not to the 500 Hz emitted by their own wings and those of neighbouring males. This frequency of 500 Hz does excite the male antennae to vibrate, but only to a very small extent. As soon, however, as a female flies nearby, the male antennae start to vibrate in resonance, that is with a considerable amplitude, even though the 300 Hz sound itself may be very faint. Johnston's organ reports this vibration to the brain, the flight mechanism is orientated precisely in the direction from which the sound is coming, and off go the male flies. In this way the males are able to pick out the females almost automatically, and in spite of the proximity of thousands of other males in the swarm, because each male receiver is permanently tuned to the female transmitter.

An exactly similar resonance effect is familiar to us all. If we turn the tuning knob of the radio to 'Munich 800 kHz', for example, this turns a tuning condenser, which is linked with other components to form a tuning circuit. As soon as the circuit is tuned exactly to to 800 kHz it resonates strongly, Munich is now tuned in, and the programme can be heard. If a neighbouring station is transmitting on 500 Hz at the same time, the signal is not heard, even if it is operating at ten thousand times as much power as the first. This transmitter is not tuned in, because our receiver is not in resonance with it. Tuning can be carried out so precisely that it is possible to take our receiver to the very foot of the Munich radio tower and still tune in to hear music from some other station, eg Moscow. In this position of the receiver the difference in strength between the two signals may be as much as four million to one, but our equipment is in tune with Moscow and not with the adjacent Munich transmitter.

Admittedly, the tuning ability of male mosquitoes is not quite as precise as this, but it is still impressive enough.

This brief summary of resonance may be rounded off with a comment by Martin Lindauer:

'Whereas male mosquitoes are attracted by a female flight tone that is different for each species, but always lower in pitch than that of the corresponding male, in olive flies it is the females that are attracted by the flight tone of the males. The males gather in groups on vegetation and send out signals resembling Morse code by vibrating their wings.' which only males have.'

Thus nothing in nature is unique. Every example of one kind can be matched by its opposite.

43. Aerial combat at night, with supersonic direction-finding and 'sonar jamming'

Bats orientate themselves in the dark by using supersonic vibrations. A bat can be confined in a dark room with fine threads stretched across it, and will fly about among these just as well as if they were fully visible, never colliding with any of them. The bat's optical system is only indifferent—the eyes are small, and in some species seem to be actually degenerate—while its audio direction finding is quite remarkable.

It works like radio equipment. An impulse is sent out and any obstacle in its path sends back an echo, which tells the bat that there is something in its way. The transmitted impulse is not on the decimetre waveband, however, but is a sound wave, and so the bat does not have a radar but a sonar.

If a sufficient number of sound impulses follow one

another at short intervals, and so send back many echoes from many objects in the vicinity, the bat can construct for itself a precise 'sound picture' of its surroundings, and this may be no less clear than the visual picture that many animals can see. It is quite outside our own experience, however. The impulses sent out by different species of bat may be either short, sharp noises, or very high notes of longer duration. They may be radiated either from the mouth or from the nasal cavities. The two ears act as direction-finding equipment for detecting the echoes. Such signals may be sent out at the rate of several dozen per second, but the rate of transmission is usually less than this.

Many bats transmit an almost pure tone of somewhat longer duration but the frequency is so high that human beings can no longer hear it. Our perception of sounds comes to an end round about 20 kHz at the maximum, but bats, on the other hand, are just beginning at this point. They often transmit at frequencies between 50 and 80 kHz, and sometimes far above this. Then the wavelength becomes so short that even the tiniest objects that might serve as prey for the bat, such as moths, or even vinegar flies *(Drosophila),* throw back a clear echo. The brain of the bat is able to make instantaneous calculations of the size, distance away, speed and direction of flight of any night-flying insect that happens to pass within the range of the bat's transmission. Immediately, the bat goes into action like a night fighter, pursues its prey and ends up with its mouth closing neatly over the victim. The vinegar fly *(Drosophila),* is often used as prey in experimental work with bats and is usually caught because it has no means of resisting capture. Sometimes the bat is not quite accurate enough in its attack, and then it does a backstroke with its wings, bends its body into an arc, and so manages to get the fly into its mouth after all.

On the the other hand, bats often have trouble in catching night-flying Lepidoptera. Suppose a bat has located a little moth and launched an attack on it. Now comes a bitter air-battle, a nocturnal life-and-death pursuit. Pursuer and pursued is each equipped with the most sophisticated equipment for its own particular purpose, and so ensues a skirmish that seems soundless to us. The bat tries to keep track of the moth, and to catch up with it, but the moth has an organ with which it can detect the sonar of the bat, up to a frequency of 200 kHz. The moth can work out from what direction the sound is coming, and tries to avoid the bat by sudden swerves, or simply by letting itself fall out of the sky. Many night-flying moths, for example the tiger moths *(Arctiidae)* can even bring a transmitter of their own into action, with which they can possibly confuse the direction-finding equipment of the bat.

Furthermore, the bodies of moths are covered with a soft, fluffy layer of scales, which absorbs a good deal of the energy

Fig. 57. The hearing organ (Johnston's organ) in the antenna of the mosquito Anopheles stephensi (Liston). A 3-dimensional reconstruction of scape, pedicel and base of the flagellum. The flagellum has a platform-like expansion at its base (Pl), which is suspended in the centre of the globular pedicel among an array of the finest struts (Sp), ribs (R) and septa (Spt). The flagellum may be set in vibration either by air currents falling on the hairs (H), or by sound waves. A concentric sclerotised ring (SkLR) in the upper cavity of the pedicel acts as a damper and prevents the vibration from building up into too great an amplitude. Three groups of mechano-sensitive cells, the outer (ScA) and the inner (ScB) rings of scolopidia, and the single, plate-like scolopidium (ScC) act by converting the kinetic energy of the vibration into electrical energy which is passed through the nerve complex of Johnston's organ (NJ) to the brain. The cross-section shows the cell nuclei (SZ), the pencils of scolopidia (St) and the terminal fibres of the sensory cells (F). The base of the pedicel extends through the scape so that the attached muscles can turn the antennae in all directions, though not very far. Further lettering: G, blood vessel; NFl, flagellar nerves.

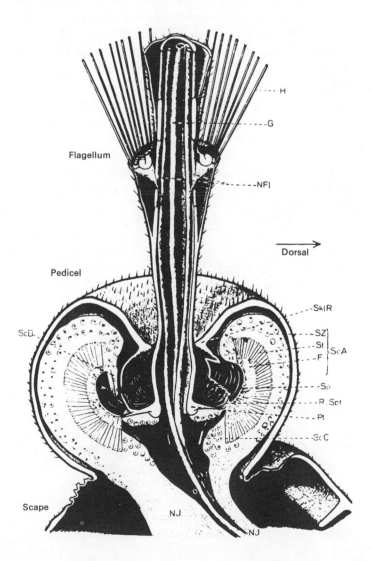

of the sonar impulse, and as a result the echo is faint and indistinct. Finally, the bat hears not only its own sonar echo, but also the normal sound produced by the flying of its prey. Moths and other night-flying Lepidoptera, however, have developed an extremely noiseless flight to reduce their own tell-tale signal to a minimum.

So the fierce struggle goes on, surging to and fro, up and down, in tight spirals and steep turns: transmitter and counter-transmitter; sonar and sonar jamming; flight manoeuvre and counter-manoeuvre; tracking by the bat, deliberate or random changes of direction by the moth; the bat's equipment trying always to obtain the clearest echo, the moth's equipment aiming always to obscure it as much as possible; the hunter listening for flight-noises, the hunted trying to smother them. This is a battle of one technology against another.

Aerial battles like this can often be watched in the light of a street-lamp, with everything happening at high speed, without concealment but in ghostly silence. If a powerful floodlight is used in place of the street-lamp, and the camera shutter is left open for a time-exposure, the paths of pursuer and pursued can be recorded on film. Fig. 58 shows two examples. In the upper picture the narrow, jagged track of the prey can be seen to intersect the broader track of the bat, and then to terminate suddenly. This is not because the prey has disappeared, but because the bat has eaten it. In the lower picture, however, it can be seen how the prey carried out a brisk evasive manoeuvre upwards, and thereby tricked its pursuer completely.

The evasive moth possesses two small sense organs, one on each side of the body, with which it can detect the ultra-sonic transmissions of the bat. Whereas sense organs commonly consist of up to ten thousand cells, these two ultrasonic detectors are only two tiny cells!

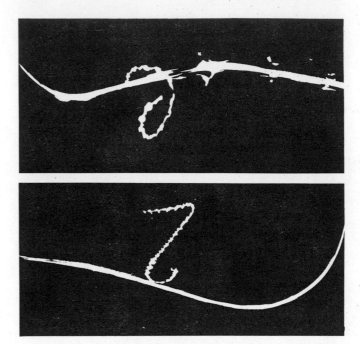

Fig. 58. A bat in combat with a night-flying moth. The brightly illuminated bat flew from right to left, and on a time-exposure produced a broad white streak on the film. Here and there can be seen individual impressions of the wings when they happened to be in such a position that they caught the light. Three can be seen on the upper right. The moth responded to the transmission of the bat by taking sharp evasive action. Each swerve of the moth produced a broadening of the path of the bat. In the upper picture the track of the moth comes to an abrupt end in contact with that of the bat: the moth was caught and eaten. In the lower picture the moth escaped upwards.

44. Comparison between insect and human flight

The newspapers occasionally draw comparisons between man's work and nature's, and there was once a time when the pioneers of human flight were struggling along in the wake of nature. As recently as the end of the nineteenth century, flying-machines were built in the shape of gigantic birds. Several times in this book I have drawn similar comparisons between man and nature, but I have always been careful to emphasise that such juxtapositions have their problems. Nature does not present us with blueprints from which technologists can borrow indiscriminately.

It is not feasible just to build a monster insect into which man can climb, and fly. The main reason is that the quantities that are of practical importance—such as rigidity, elasticity, moment of inertia, wind force, fuel requirement, wing loading, atmospheric pressure and a multiplicity of other parameters—change in the most varied ways if the size and velocity of an object are changed. We have already seen, when we were considering gliding flight, that the gliding angle of a flying object is a built-in characteristic, and that its value changes if one and the same design of flying-machine is constructed on a different scale. One of the factors may perhaps change in linear fashion as the length of the flying-machine is increased, while another may increase with the square of the length, a third with the cube, a fourth may be proportional to the square root of the length, and a fifth may vary exponentially.

If it were possible to make an exact replica of a fly, but a thousand times bigger, we should then have a giant insect 12 metres long, made out of aluminium and other light materials. If it were set down on the runway it might look like a normal aeroplane, but it would never fly, at least in any way resembling a real insect.

For one thing, nature uses quite different building materials from those used by man. She uses vast quantities of elastic 'synthetics', sometimes, as with resilin, even substances of extremely idiosyncratic properties. When man builds an aeroplane, everything has to have stiffness, if not rigidity, whereas in nature nothing is so: everything is pliant, elastic, capable of being deformed. Nature has her own building laws, which work well at these tiny dimensions, but break down when raised to the scale of human constructions.

This is not to say, however, that man cannot make use of materials of such a kind, and that flight by means of flapping wings is outside the bounds of possibility from the start. Quite the reverse. In spite of the many attempts that have been made, some of them well considered, but most of them less so, we are still only on the threshold of this field.

Many people have wished that technology could invent some mechanical principle that would be sparing of energy, and economical of fuel. For certain special requirements, such as slow flight combined with extreme manoeuvrability for example, this would be better than either the propellor or the jet as a driving force. The engineer could thus learn a great deal by studying how nature has solved the many technical problems of flight, but he cannot take nature's solutions and use them without adaptation. He must behave as a good engineer should; observe and compare, pick up ideas, but not copy them slavishly, work upon them, and use them as starting points for his own constructions. The branch of research called 'bionics' (an American abbreviation for 'biotechnology') has precisely this for its objective. Research workers in bionics eavesdrop on the secrets of nature and adapt these for their own purposes. They have

already used such 'stolen' ideas to invent mechanisms for orientation, heat detectors, velocity meters, zip-fasteners and every other possible kind of device.

Up until now technology has not made very much use of the principles of flight as the animals use them, but there has been no lack of suggestion. One example must serve to represent many. Towards the end of the Second World War the Focke–Wulff Company had a plan to build an airliner with two pairs of wings, beating alternately like those of a dragonfly. The upper drawings in Fig. 59, reading from right to left, show successive developments of a flying-machine based upon such an idea taken from an insect, and demonstrate very well that the idea is mechanically sound. The principle was altered, however, during the course of its development, to bring it more into line with technological possibilities: six propelling surfaces instead of four, contra-rotation in place of alternating up and down movements. Tailplane and vertical fin replaced the plain, cylindrical abdomen of the dragonfly, since it was technically difficult to control the aeroplane by flexing the wing surfaces (though this was done in the earliest man-made aeroplanes).

Erich von Holst, a biologist, suggested this idea, and based it upon a single biological model, the dragonfly. A technologist might have arrived at the ideal solution from

Fig. 59. The sequence of ideas by which a biologist (above) and a technologist (below) might each arrive at the idea of a contrarotating propeller. Each sequence should be read from right to left, and is explained in the text.

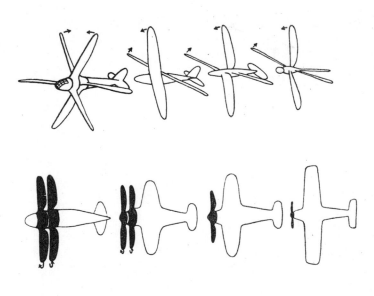

mechanical principles, without depending in any way upon a natural model. He, of course, would start from a man-made object, an ordinary airliner, and his train of thought is represented by the four sketches in the lower row of Fig. 59, also from right to left. To achieve a high performance from either a natural or a man-made starting-point, the same basic principle is involed: to balance the moment of inertia of a single propelling surface with the reaction of one that is moving in the opposite direction. The idea can either be taken direct from nature, or it can be worked out afresh if the problem is thought over long enough.

If you ask me what are the greatest benefits that the flight technologist might derive from the study of insect flight, my answer would be: 'the dynamics of propulsion for the development of the helicopter; the art of combining turning and twisting movements for the development of long-range, slow-flying, wing-flapping aircraft; perhaps static theory and constructional ideas which would contribute to the same purpose; some constructional material resembling resilin as an energy accumulator; a retractable contrivance like a set of legs to give a jump-start; the thoracic structure of a dragon-fly to suggest ideas for cabin construction in light aircraft; and the basic principles of various measuring and monitoring devices'. Who knows, however, whether some quite different aspect of the subject might not become particularly important in future, some small detail that at this present moment seems quite insignificant to us.

Who can possibly forecast the future development of this fascinating branch of modern biophysics, of which I have been able to reveal just a little in this book? Anyway, it is certain that the biophysics of animal flight still has a great deal to teach us.

Bibliography

Chapter 1.

FORSTER, W. *Knaur's Insektenbuch*. Droemer-Knaur, München–Zürich (1968)
HESSE, H. and DOFLEIN, F. *Tierbau und Tierleben, in ihrem Zusammenhang betrachtet*. Vol. 1 and 2, G. Fischer, Jena (1935–1943)
KAESTNER, A. *Lehrbuch der Speziellen Zoologie: Insekten*. G. Fischer, Stuttgart (1970)
v. KÉLÉR, S. *Entomologisches Wörterbuch*, 1st edn., Akademie-Verlag, Berlin (1955)
v. LENGERKEN, H. *Insekten (Sammlung 'Das Tierreich')*, Vol. IV/3, Walter de Gruyter, Berlin, 2nd edn.
SNODGRASS, R. E. *Anatomy and Physiology of the Honeybee*. McGraw Hill, New York (1925)
WEBER, H. *Grundriß der Insektenkunde*. 2nd edn., G. Fischer, Jena (1949)

Chapter 2.

BROHMER, P. (ed.) *Die Tierwelt Mitteleuropas*, Vols. IV, V and VI: *Insekten*. Quelle & Meyer, Leipzig
BROHMER, P. *Fauna von Deutschland*. 9th edn., Quelle & Meyer, Heidelberg (1964)
BRONN, H. G. *Klassen und Ordnungen des Tierreichs*. Vol. V, part III: *Insecta*. Akad. Verlagsges. Geest & Portig, Leipzig (1955)
DOEDERLEIN, L. *Bestimmungsbuch für deutsche Land- und Süßwassertiere. Insekten*, part I, 2nd edn., revised by W. JACOBS. Oldenbourg, München (1952)
KLOTS, A. and KLOTS, E. B. *Knaur's Tierreich in Farben: Insekten*. Droemer-Knaur, München–Zürich (1959)
STRESEMANN, E. *Exkursionsfauna*. Volk-und-Wissen-Verlag, Berlin

Chapter 3.

COMSTOCK, J. H. *The Wings of Insects*. Ithaca (1918)
GRASSÉ, P.-P. *Traité de Zoologie*, Vol. IX–X: *Insectes*. Masson & Cie, Paris (1949–1951)
HERTEL, H. *Biologie und Technik*. Krausskopf, Mainz (1963)
KLEINOW, W. Untersuchungen zum Flügelmechanismus der Dermapteren. *Z. Morphol. Ökol. Tiere* 56, 363–416 (1966)
SCHMIDT, H. *Der Flug der Tiere*. Kramer, Frankfurt (1960)
SNODGRASS, R. E. *Principles of Insect Morphology*, chapter X, The Wings. McGraw Hill, New York (1935)

Chapter 4.

NACHTIGALL, W. Aerodynamische Messungen am Tragflügelsystem segelnder Schmetterlinge. *Z. vergl. Physiol.* 54, 210–31 (1967)
MAGNAN, A. and GIRERD, H. Sur la détermination en soufflerie des polaires des papillons, *C. R. Acad. Sci.*, 198, 243 (1934)

Chapter 5.

GLADKOW, N. A. *Flüge in der Natur*. Urania, Jena (1953)
GRAY, J. *How Animals Move*. Pelican Books, Edinburgh (1957)
KATZ, G. *Das kleine Buch vom Papierflugzeug*. Fretz & Wasmuth, Zürich (1953)

Chapter 6.

BERGMANN, L. and SCHAEFER, C. *Lehrbuch der Experimentalphysik*, Vol. I. Walter de Gruyter, 2/3 edn., Berlin (1945)

Chapter 7.

GYMNICH, A. *Der Segelflug in Theorie und Praxis*. Otto Maier, Ravensburg, 2nd edn. (1957)
SCHMARSOW, G. *Die Physik des Fliegens*. Fachbuchverlag Leipzig (1955)
SCHMITZ, F. W. *Aerodynamik des Flugmodells*. 2nd edn. Duisburg (1952)
SCHÜTT, K. *Einführung in die Physik des Fliegens*. 2nd edn., Volkmann, Berlin (1934)

Chapter 8.

BURTON, A. J. and SADEMAN, D. C. The lift provided by the elytra of the Rhinoceros beetle, *Oryctes Boas Fabr. South Afric. J. Sci.* 57, (No. 4) 107–9 (1961)
NACHTIGALL, W. Zur Aerodynamik des Coleopterenfluges: Wirken die Elytren als Tragflügel? *Verh. deutsch. Zool. Ges.* (Kiel) 58, 319–26 (1964)
NACHTIGALL, W. Die aerodynamische Funktion der Schmetterlingsschuppen. *Naturwiss.* 52, 216–17 (1965)
NACHTIGALL, W. Aerodynamische Messungen am Tragflügelsystem segelnder Schmetterlinge. *Z. verg. Physiol.* 54, 210–31 (1967)
VOGEL, S. Flight in *Drosophila*, III: aerodynamic characteristics of fly wings and wing models. *J. Exp. Biol.* 46, 431–43 (1967)

Chapter 9.

JONES, G. A. *High Speed Photography*. London (1952)

Chapter 10–15.

BETTS, A. How bees fly. *The Bee World* 14, 50–5 (1933)
CHADWICK, L. The wing motion of the dragonfly. *Bull. Brooklyn Entomol. Soc.* XXXV (No. 4), 109–12 (1940)
HOLLICK, F. S. J. The flight of the dipterous fly *Mucina stabulans* Fallén. *Phil. Trans. Roy. Soc. Lond. B*, 230, 357–90 (1940)
MAGNAN, A. *Le Vol des Insectes*. Herrmann & Cie, Paris (1934)
NACHTIGALL, W. Die Kinematik der Schlagflügelbewegungen von Dipteren. Methodische und analytische Grundlagen zur Biophysik des Insektenflugs. *Z. vergl. Physiol.* 50, 149–211 (1966)
PRASSE, J. Über den Start und Flugs des *Sisyphus schaefferi* L. *Beitr. Entomol.* 10 (1/2), 168–83 (1960)
PRINGLE, J. W. S. *Insect Flight*. London, Cambridge University Press (1957)
PRINGLE, J. W. S. *Locomotion: Flight*. In Roeder's *The Physiology of Insecta*, Vol. II, pp. 283–327, Academic Press, New York–London (1965)
PRINGLE, J. W. S. Comparative Physiology of the Flight Motor (in Press).
VOGEL, S. Flight in *Drosophila*, I: flight performance of tethered flies. *J. Exp. Biol.* 44, 567–78 (1966)
VOGEL, S. Flight in *Drosophila*, II: variations in stroke parameters and wing contour. *J. Exp. Biol.* 46, 383–92 (1967)
WEIS-FOGH, T. and JENSEN, M. Biology and physics of locust flight, II: flight peformance of the desert locust (*Schistocerca gregaria*). *Phil. Trans. Roy. Soc. Lond. B*, 239, 459–510 (1956)
WOHLGEMUTH, R. Die Schlagform des Bienenflügels beim Sterzeln im Vergleich zur Bewegungsweise beim Fliegen und Fächeln. *Z. vergl. Physiol.* 45, 581–89 (1962)

Chapter 16.

BENNETT, L. Insect aerodynamics: vertical sustaining force in near-hovering flight. *Science* 152, 1263–6 (1966)
DUBS, F. *Aerodynamik der reinen Unterschallströmung*. Birkhäuser, Basel–Stuttgart (1954)
v. HOLST, E. Untersuchungen über Flugbiophysik I. Messungen zur Aerodynamik kleiner schwingender Flügel. *Biol. Zbl.* 63, 289–326 (1943)
JENSEN, M. Biology and physics of locust flight III: the aerodynamics of locust flight.

Phil. Trans. Roy. Soc. Lond. B, **239**, 511–52 (1956)

NACHTIGALL, W. Die Kinematik der Schlagflügelbewegungen von Dipteren etc. *Z. vergl. Physiol.* **50**, 149–211 (1966)

OSBORNE, M. F. M. Aerodynamics of flapping flight with application to insects. *J. Exp. Biol.* **28**, 221–45 (1951)

SEMJONOW, V. *Einführung in die Flugphysik,* Part II. Schmidt & Co, Braunschweig (1955)

WEIS-FOGH, T. Biology and physics of locust flight, I: basic principles in insect flight. A critical review. *Phil Trans. Roy. Soc., Lond. B,* **239**, 415–58 (1956)

Chapter 17.

FABRE, J. H. *Aus der Wunderwelt der Insekten.* Hain, Mesenheim (1950)

SCHREMMER, F. Gezielter Abwurf getarnter Eier bei Wollschwebern. *Verh. deutsch. Zoolog. Ges* (München 1963) 291–303 (1964)

WESENBERG-LUND, C. *Biologie der Süßwasserinsekten.* Springer, Berlin (1943)

Graf ZEDWITZ, F. *Wunderbare kleine Welt.* Safari, Berlin (1952)

ZSCHOKKE, F. *Der Flug der Tiere.* Springer, Berlin (1919)

Chapter 18.

GREENWALT, C. H. The wings of insects and birds as mechanical oscillators. *Proc. Amer. Phil. Soc.* **104**, 605–11 (1960)

HOCKING, B. Insect flight. *Sci. Amer.* **199**, 92–8 (1958)

SMITH, D. S. The flight muscles of insects. *Sci. Amer.* **212**, 76–88 (1965)

Chapter 19.

BOETTIGER, E. G. and FURSHPAN, E. The mechanics of flight movements in diptera. *Biol. Bull.* **102**, 200–11 (1952)

MIHÁLYI, F. Untersuchungen über Anatomie und Mechanik der Flugorgane an der Stubenfliege. *Arb. Ungar. Biol. Forschungsinstitut Tihany* **8**, 106–119 (1935/36)

Chapter 20.

DAVIS, R. A. and FRAENKEL, G. The oxygen consumption of flies during flight. *J. Exp. Biol.* **17**, 402–7 (1940)

HILDEBRANDT, A. *Luftschiffahrt.* Oldenbourg, München, Berlin (1910)

LUDWIG, H. W. O_2-Konsum und RQ fliegender Insekten. *Verh. deutsch. Zool. Ges.* (Wien 1962) 355–9 (1963)

MILLER, P. L. Respiration in the desert locust, III: ventilation and the spiracles during flight. *J. Exp. Biol.* **37**, 264–78 (1960)

WEIS-FOGH, T. Functional design of the tracheal system of flying insects as compared with the avian lung. *J. Exp. Biol.* **41**, 207–27 (1964)

Chapter 21.

CLEGG, J. S. and EVANS, D. R. The physiology of blood trehalose and its function during flight in the blowfly. *J. Exp. Biol.* **38**, 771–92 (1961)

GREEN, D. E. The Mitochondrion. *Sci. Amer.,* Sonderband 'The Living Cell', 63–71 (1965)

NEVILLE, A. C. Energy and economy in insect flight. *Sci. Progr.* **53**, 203–19 (1965)

WEISS-FOGH, T. Fat combustion and metabolic rate of flying locusts *(Schistocerca gregaria,* Forskål) *Phil. Trans. Roy. Soc. Lond. B,* **237**, 1–36 (1952)

Chapter 22.

ADAMS, P. A. and HEATH, J. E. Temperature regulation in the sphinx moth, *Celerio lineata, Nature* **201**, 20–2 (1964)

ESCH, H. Über die Körpertemperaturen und den Wärmehaushalt von *Apis mellifica. Z. vergl. Physiol.* **43**, 305–35 (1960)

SOTAVALTA, O. On the thoracic temperature of insects in flight. *Ann. Zool. Soc. Fenn. 'Vanamo'* **16**, 1–21 (1954)

Chapter 23.

ANDERSEN, S. O. and T. WEIS-FOGH. Resilin. A rubberlike protein in arthropod cuticle. *Adv. Insect Physiol.* **2**, 1–65

Chapter 24.

RUSSENBERGER, H. and RUSSENBERGER, M. Bau und Wirkungsweise des Flugapparats von Libellen, mit besonderer Berücksichtigung von Aeschna cyanea. *Mitt. Naturforsch. Ges. Schaffhausen* **27**, 1–88 (+12 Tafeln) (1959/1960)

SNODGRASS, R. E. *Principles of Insect Morphology.* Ch. VIII: The Thorax. McGraw Hill, New York–London (1935)

Chapter 25.

DEMOLL, R. *Der Flug der Insekten und der Vögel.* G. Fischer, Jena (1918)

HOCKING, B. The intrinsic range and speed of flight of insects. *Trans. Roy. Soc. Ent. Soc.* **104**, 223–42 (1939)

LANGMUIR, J. The speed of the deer fly. *Science* **87**, 233–4 (1938)

Chapter 26.

HORRIDGE, G. A. The flight of very small insects. *Nature* **178**, 1334–1335 (1956)

NACHTIGALL, W. Funktionelle Morphologie, Kinematik und Hydromechanik des Ruderapparats von *Gyrinus. Z. vergl. Physiol.* **45**, 193–226 (1961)

SLIJPER, E. J. *Riesen und Zwerge in Tierreich.* Parey, Hamburg (1967)

THOM, A. and SWART, P. The forces on an aerofoil at very low speeds. *J. Roy. Aero. Soc.* **44**, 761–70 (1940)

Chapter 27.

HERTEL, H. *Biologie und Technik.* Krausskopf, Frankfurt (1963)

Chapter 28.

BULLOCK, T. H. and HORRIDGE, G. A. *Structure and Function in the Nervous System of Invertebrates.* Vol. I and II. Freeman and Co, San Francisco/London (1965)

WHITFIELD, J. C. *Electronics for Physiological Workers.* 2nd edn., Macmillan, London/New York (1959)

Chapter 29.

WALDRON, J. Neural mechanism by which controlling inputs influence motor output in the flying locust. *J. Exp. Biol.* **47**, 213–28 (1967)

WILSON, D. M. The central nervous control of flight in a locust. *J. Exp. Biol.* **38**, 471–90 (1961)

WILSON, D. M. Phasically unpatterned nervous control of dipteran flight. *J. Ins. Physiol.* **9**, 859–65 (1963)

WILSON. D. M. The nervous co-ordination of insect locomotion. In *The Physiology of the Insect Central Nervous System.* Academic Press, London/New York

WYMAN, R. J. Multistable firing patterns among several neurons. *J. Neurophysiol* **29**, 807–33 (1966)

Chapter 30.

GREEN, D. E. and HATEFI, Y. The mitochondrion and biochemical machines. *Science* **133**, 13–19 (1961)

HUXLEY, H. E. The contraction of muscle. *Sci. Amer.,* Sonderband 'The Living Cell', 278–89 (1965)

PRINGLE, J. W. S. The contractile mechanism of insect fibrillar muscle. *Progr. Biophys. Molec. Biol.* **17**, 1–60 (1967)

SMITH, D. S. The flight muscles of insects. *Sci. Amer.* **212**, 76–88 (1965)

Chapter 31.

NACHTIGALL, W. and WILSON, D. M. Neuromuscular control of dipteran flight. *J. Exp. Biol.* **47**, 77–97 (1967)

WILSON, D. M. Neurale Kontrolle des Heuschrekkenfluges. *Umschau,* **24**, 817 (1966)

Chapter 32.

DUGARD, J. J. Directional change in flying locusts. *J. Ins. Physiol.* **13**, 1055–63 (1967)

NACHTIGALL, W. and WILSON, D. M.Neuromuscular control of dipteran flight. *J. Exp. Biol.* **47,** 77–97 (1967)

Chapter 33.
GOODMAN, L. J. The landing responses of insects. I. The landing response of the fly, *Lucilia sericata,* and other Calliphorinae. *J. Exp. Biol.* **37,** 854–78 (1960)

Chapter 34.
KAMMER, A. E. Phase relationships between motor units during flight and preparation for flight in hawk moths. *Amer. Zoologist* **5,** No. 4, (1965)
NACHTIGALL, W. and WILSON, D. M. Neuromuscular control of dipteran flight. *J. Exp. Biol.* **47,** 77–97 (1967)
WALDRON, J. Mechanisms for the production of the motor output pattern in flying locusts. *J. Exp. Biol.* **47,** 201–12 (1967)

Chapter 35.
CURRAN, C. H. How flies fly. *Natural History* **67,** 82–5 (1958)
HYZER, W. G. Flight behavior of a fly alighting on a ceiling. *Science* **137,** (No. 3530), 24. (1962)
LINDNER, E. Die Stubenfliege an der Zimmerdecke. *Kosmos* **55,** 54–5 (1959)

Chapter 36.
DADE, H. A. The flight of the honeybee. *Bee World* **43** (No 1), 12–20 (1962)
KAUFMANN, T. Faltungsmechanismen der Flügel bei einigen Coleopteren. *Diss. Nat. Fak. Univ. München* (1960), unpubl.

Chapter 37.
JOHNSON, C. G. The aerial migration of insects. *Sci. Amer.,* Dec. (1963)
UVAROV, B. *Grasshoppers and Locusts.* Vol. I. London, Cambridge University Press. (1966)

Chapter 38.
NIELSEN, E. T. *Insekten auf Reisen.* Series 'Verständliche Wissenschaft', Springer, Berlin/Heidelberg–New York (1967)
WILLIAMS, C. B. *Insect Migration.* Collins (1958)

Chapter 39.
ESCH, H. Beiträge zum Problem der Entfernungsweisung in den Schwänzeltänzen der Honigbiene. *Z. vergl. Physiol.* **48,** 543–6 (1964)
ESCH, H. The Evolution of Bee Language. *Sci.*

Amer. **216,** (No. 4), 96–104 (1967)
v. FRISCH, K. Die Tänze der Bienen. *Österr. Zool. Z.* **1,** 1–48 (1946)
v. FRISCH, K. *Sprache und Orientierung der Bienen.* Memorial lecture for Dr Albert Wander. Vol. 3, H. Huber, Bern/Stuttgart (1961)
v. FRISCH, K. and KRATZKY, O. Über die Beziehung zwischen Flugweite und Tanztempo bei der Entfernungsmeldung der Bienen. *Naturwiss.* **49,** 409–17 (1962)
LINDAUER, M. Allgemeine Sinnesphysiologie. Orientierung im Raum. *Fortschr. Zool.* **16,** 58–140 (1963)
SCHOLZE, E., PICHLER, H. and HERAN, H. Zur Entfernungsschätzung der Biene nach dem Kraftaufwand. *Naturwiss* **51,** 69–70 (1964)
WENNER, A. M. Sound production during the waggling dance of the honeybee. *Anim. Behav.* **10,** 79–95 (1962)

Chapter 40.
BURKHARDT, D. and SCHNEIDER, G. Die Antennen von *Calliphora* als Anzeiger der Fluggeschwindigkeit. *Z. Naturforschung* **12b,** 139–43 (1957)
GEWECKE, M. Die Wirkung von Luftströmung auf die Antennen und das Flugverhalten der blauen Schmeißfliege *(Calliphora erythrocephala). Z. vergl. Physiol.* **54,** 121–64 (1967)
HERAN, H. Wahrnehmung und Regelung der Flugeigengeschwindigkeit bei *Apis mellifica* L. *Z. vergl. Physiol.* **42,** 103–63 (1959)
HERAN, H. Wie überwacht die Biene ihren Flug? *Umschau* **64,** Part 10, 299–303 (1964)
HERAN, H. and LINDNER, M. Windkompensation und Seitenwindkorrektur der Bienen beim Flug über Wasser. *Z. vergl. Physiol.* **47,** 39–55 (1963)

Chapter 41.
FAUST, R. Untersuchungen zum Halterenproblem. *Zool. Jb.* **63,** 325–66 (1952)
FRAENKEL, G. and PRINGLE, J. W. S. Halteres of flies as gyroscopic organs of equilibrium. *Nature* **141,** 919–20 (1938)
GETTRUP, E. Sensory mechanisms in locomotion: the campaniform sensilla of the insect wing and their function during flight. *Cold Spring Harbor Symp. Quant. Biol.* XXX, 615–22 (1965)
GETTRUP, E. and WILSON, D. M. The lift control reaction of flying locusts. *J. Exp. Biol.* **41,** 183–90 (1964)
PABST, H. Elektrophysiologische Untersuchungen des Streckrezeptors am Flügel-

gelenk der Wanderheuschrecke *Locusta migratoria. Z. vergl. Physiol.* **50,** 498–541 (1965)
PRINGLE, J. W. S. The gyroscopic mechanism of the halteres of diptera. *Phil Trans. Roy. Soc. Lond. B,* **233,** 347–84 (1948)
WEIS-FOGH, T. Biology and physics of locust flight IV: notes on sensory mechanisms in locust flight. *Phil. Trans. Roy. Soc. Lond. B,* **239,** 553–84 (1956)

Chapter 42
RISLER, H. Das Gehörörgan der Männchen von *Anopheles stephensi* Liston (Culicidae). *Zool. Jb. Anat.* **73,** 165–85 (1953/54)
SOTAVALTA, O. The flight-tone of insects. *Acta Entomol. Fenn.* **4,** 1–115 (1947)
TISCHNER, H. Gehörsinn und Fluggeräusch von Stechmücken. *Umschau* **12,** 368–70 (1955)
TISCHNER, H. Ortungsbiologie. *Bild der Wissenschaft,* February 1965, pp. 138–47
WESENBERG-LUND, C. *Biologie der Süßwasserinsekten.* Springer, Berlin (1943)
WISHART, G. and RIORDAN, D. F. Flight response to various sounds by adult males of *Aedes aegypti* (L) (Diptera: Culicidae). *The Canadian Entomologist* XCI, 181–91, March 1959

Chapter 43.
GRIFFIN, D. R. Echo-Ortung der Fledermäuse, insbesondere beim Fangen fliegender Insekten. *Naturwiss. Rundschau* **15,** 169–73 (1962)
ROEDER, K. D. Moths and ultrasound. *Sci. Amer.* **212,** 94–102 (1965)
ROEDER, K. D. and PAYNE, R. S. Acoustic orientation of a moth in flight by means of two sense cells. *Symp. Soc. Exp. Biol.* **20,** 251–72 (1965)
ROEDER, K. D. Acoustic sensitivity of the noctuid tympanic organ and its range for the cries of bats. *J. Ins. Physiol.* **12,** 843–59 (1966)
WEBSTER, F. A. The role of the flight membranes in insect capture by bats. *Anim. Behav.* **10,** 332–40 (1962)

Chapter 44
FITZ-PATRICK, J. L. G. *Natural Flight and Related Aeronautics.* Sherman M. Fairchild Publication Fund Paper, New York (1952)
v. HOLST, E. Prinzipien des Tierflugs und ihre technische Bedeutung. *Fra Fysikkens Verden,* Part 2, 53–68 (1961)
v. HOLST, E. and KÜCHEMANN, D. Der Triebflügel. *Jb. deutsch. Luftfahrtforschung,* 435–43 (1942)

1. Adapted from different illustrations in *Traité de Zoologie*, Vols. IX–X (1949–51)
2. v. Keler, S. (1955) Plate VI, Figure 1 and the figure on Plate XXIX
3. Adapted from different illustrations in *Traité de Zoologie*, Vols. IX–X (1949–51)
4. Kleinow, W. (1966), Fig. 1, p. 364
5. Weber, H. (1949), Fig. 42, p. 50 (altered)
6. Weber, H. (1949), Fig. 177, p. 227 (part)
7. *Traité de Zoologie* (1951), Vol. X, Fig. 229, p. 204
8. *Traité de Zoologie* (1951), Vol. X, Figs. 996 and 529, pp. 1099 and 539
9. Gray, J. (1957), Fig. 49, p. 130 (redrawn)
10. Nachtigall, W. (1968), original
11. Nachtigall, W. (1968), original
12. Nachtigall, W. (1968), original
13. Nachtigall, W. (1968), original
14. Nachtigall, W. (1967), Fig. 13, p. 229 (part)
15. Nachtigall, W. (1967), Fig. 10, p. 226
16. Vogel, S. (1967), Fig. 7, p. 439 (altered)
17. Nachtigall, W. (1968), original
18. Nachtigall, W. (1966), Fig. 3, p. 161 (altered)
19. Nachtigall, W. (1966), Fig. 10, p. 170 (altered)
20. Nachtigall, W. (1966), Fig. 7, p. 166
21. Nachtigall, W. (1966) Fig. 17, p. 181
22. Nachtigall, W. (1966), Fig. 18, p. 183 (altered)
23. Nachtigall, W. (1966), Fig. 25, p. 193 (altered)
24. Nachtigall, W. (1966), Fig. 23, p. 190
25. Nachtigall, W. (1968), original
26. Nachtigall, W. (1968), original
27. Nachtigall, W. (1966), Fig. 26, p. 194 (altered)
28. Nachtigall, W. (1968), original
29. v. Holst, E. (1951), Fig. 11, p. 63 (altered)
30. Smith, D. S. (1965), Fig. 1, p. 79 (redrawn and altered)
31. Nachtigall, W. (1968), original
32. Smith, D. S. (1965), Fig. 1, right, p. 79 (redrawn and altered)
33. Nachtigall, W. (1968), original
34. Nachtigall, W. (1968), original
35. Nachtigall, W. (1968), original
36. Russenberger, H. and Russenberger, M. (1959–60), Fig. 4, p. 13 (redrawn)
37. *Traité de Zoologie* (1951), Vol. X, Fig. 1615, p. 1827
38. Nachtigall, W. (1961), Fig. 18, p. 219 (altered)
39. Wilson, D. M. (1966), Fig. 1, p. 817
40. Waldron, I. (1967), Fig. 1, p. 214 (part)
41. Waldron, I. (1967), Fig. 3, p. 217 (part)
42. Nachtigall, W. and Wilson, D. M. (1967), Fig. 50, p. 85
43. Wyman, R. J. (1966), Fig. 7, p. 822 (part)
44. Huxley, H. E. (1958), the figure on p. 280
45. Nachtigall, W. and Wilson, D. M. (1967), Fig. 5a, p. 84 (altered)
46. Goodman, L. J. (1960), Fig. 3, p. 859
47. Hyzer, W. G. (1962), Fig. 2 (altered)
48. Dade, H. A. (1962), Fig. 9, p. 19
49. Kaufmann, T. (1960), Fig. 10, p. 56 (altered)
50. Kleinow, W. (1966), Fig. 2, p. 367
51. Johnson, C. G. (1963), Fig. 3, p. 4 (line drawing, part)
52. v. Frisch, K. (1961), Fig. 3, p. 9
53. Nachtigall, W. (1968), original
54. v. Frisch, K. (1961), Fig. 6 below, p. 14
55. v. Frisch, K. (1961), Fig. 16, p. 26
56. Nachtigall, W. (1968), original
57. Risler, H. (1953/54), Fig. 4, p. 169 (altered)
58. Webster, F. and Roeder, D. (1965), Figs. 1 and 2, p. 95 (redrawn)
59. v. Holst, E. (1951), Figs. 12 and 15, pp. 65 and 67

Acknowledgements for tables

1. W. Nachtigall; 2. G. Schützenhoter; 3 above W. Kleinow; 3 below W. Kratz; 4. H. E. Edgerton; 5. H. E. Edgerton; 6. W. Rohdich; 7. W. Nachtigall; 8. W. Nachtigall; 9. W. Rohdich; 10. G. Olberg and F. Schremmer; 11. W. Zepf; 12. W. Nachtigall; 13. W. Kratz; 14. D. S. Smith; 15. H. E. Huxley; 16. W. Nachtigall; 17. G. Olberg; 18. H. E. Edgerton; 19. G. Olberg; 20. R. Bitschene; 21. G. Olberg; 22. G. Olberg; 23. H. Mittelstaedt, Rilling, Roeder; 24. B. Leidmann and K. Warlies; 25. Bibliothèque Nationale, Paris; 26. W. Nachtigall; 27. G. Schützenhofer; 28. © Walt Disney Productions; 29. K. Müller; 30. H. Tischner; 31. B. Leidmann and K. Warlies; 32. H. E. Edgerton; all vignettes, *Traité de Zoologie*

INDEX

acceleration, gravitational, 23
accelerator, 134–138
Acilius, 90, 91
actin, 106, 107
action potential, 76, 96–104, 117, 118
adenosine triphosphate, 106, 107
aerodynamic forces, 23–29, 35–37, 47–52
aeroplane, 26, 28, 145
Aeschna, 19, 52, 84, 88
air flow, 22, 25, 35–37, 51, 139–141
air tube, see trachae
Alucita, 62
Ammophila, 84, 88
anatomy, 9–11
Anax, 88
angle of attack, 24–29, 50, 51
angle of attack, critical, 27, 28
Anopheles stephensi, 143
ants, 120
antennae, 137–143
aphid, 129
Apis mellifera, 98, 120, 131–133
apollo butterfly, 55
Ascia monuste, 127
asynchronous flight muscles, indirect flight, see muscle
ATP, 106–108
autopilot, 141

balance, aerodynamic, 36, 45
basal hinge, 39, 61–67, 83
bat, 142–144
bee, 72, 98, 120, 131–133
bee, ground living, 55, 92
beefly, 55, 57, 92
Bembex tarsata, 56
bird, 48, 83, 90, 128, 129
bionics, 145
bistable system, see click mechanisms
blood sugar, 79
blowfly, 16, 30, 72, 77, 88, 97–98, 117–119, 134
bluebottle, see blowfly
body, ideal elastic, 82
body weight, 35–36, 79, 129
Bombylus, 55, 57
boundary layer, 29
brain, 118, 143

breathing, 69, 72, 77
bristly wings, 89–91
bumble-bee, 30, 122
butterfly, 15, 19–21, 25, 30, 51, 55, 79, 105
buzzard, 22

cabbage white butterfly, 21, 88
caddis-fly, 15
calcium, 106, 107
Calliphora, 16, 17, 65, 77, 97, 98, 109, 134
calorie, 129
camera, high speed, 30, 38, 40, 45
carbohydrate, 78, 129
carbon tetrachloride, 66, 67
catapult launch, 117, 118
CCl4, 66, 67
Celerio, 80
Celonites, 15
central nervous system, 60, 62, 65, 93, 94, 96, 103–105, 109, 110, 137, 140
centre of, mass, 23, 24, 35
cicada, 30
Cicindela, 121
classification, 12
click mechanism, 21, 65–68, 97, 103
Cocinella, 35
Coccophagus, 15
cockchafer, 88, 116, 127
comparison, biology, 25–29, 67– and technology, 69, 72, 82–84, 129, 144–146
coordinates, polar, 132, 134
Culex pipiens, 141

damping, 68, 72, 83, 89
damselfly, 88, 92
Danaus plexippus, 128
dance, bees' tail-wagging, 131–134
devil's needle dragonfly, 19
diffusion, 72
digestive system, 9, 11
direct flight muscle, 60, 96, 104, 107, 108
Dissosteira, 15
Dociostauras maroccanus, 124, 127
downstroke, 41–43, 45–49, 51, 52, 55, 56, 61, 65–68, 70, 83

drag, 25–28, 35–37, 49, 51, 56, 90, 106, 113, 118, 136, 139
dragonfly, 13, 15, 19, 21, 52, 58, 60, 77, 81, 83–85, 92, 105, 107, 120, 129, 130, 145, 146, 149
Drosophila, 28, 58, 61, 79, 83

eagle, 22
earwig, 13, 15, 30, 31, 120, 122, 123
elastic deformation, 82, 83, 88, 89
electrodes, 45, 94, 44, 101, 105, 114
electron microscope, 19, 105, 106
endoplasmic reticulum, 106, 107, 111, 112
energetics of muscle, 106–107, 128, 129
energy storing ligaments, 83, 89
endplates, 107, 111
Eoxeonos, 15
Eriosoma, 15
Eristalis, 38, 40
Esch, Harald, 132, 134
Eumenes, 1'2
Euthrips, 89, 94
evaporation, heat of, 80, 87
experimental apparatus, 36–38, 40, 44
eyes, compound, 9, 117, 128, 133, 135, 138, 139

Fabre J. H., 55, 59
fat bodies, 78, 79, 82, 83, 128, 129
feedback, negative, 37, 38, 137
film measurement, 41, 42, 45
firefly, 59, 62
flagellum, 143
flight, climbing, 109, 113, 137, 148
flight, drunken, 66, 69
flight, free, 35, 48, 79, 87, 110, 121
flight, motor, cooling, 80, 81, 87
flight muscles, 10, 65, 90, 102
    see direct flight and 105, 107
    indirect flight muscle, 111, 112
flight, normal, 35–37, 51–52, 55, 57, 83
flight, speed of, 36, 37, 84, 88, 127, 137–140, 146
flight, stabilisation, 139–141
flight, supercritical, 27–29
flight, tone, 66, 69, 95, 102, 140–144
flight, unpowered, 21, 22, 24–26
fly, flight musculature, 65

foraminifera, 38, 39
force, 23, 24, 35–37, 51, 56
force, compensating balance, 35–37
force, aerodynamic, see aerodynamic
  force
Forficula, 15, 121, 123
frequency of wing beat, 13, 60, 62, 65,
  66, 79, 91, 95–98, 103–104
fruit fly, see vinegar fly
fuel, 69, 72, 77–80, 92, 108, 131, 134,
  137–139, 146
furca, 84, 90

ganglion, 9, 11, 68, 139, 140
Gastropacha, 15
glide angle, 24–26, 146
glider, high performance, 24, 26
glide number, 24–26, 90
glide path, 24, 25
gliding, 21, 22, 24–26
glycogen, 79, 83
gnat, 22, 55, 95, 129, 141, 142
gusts of wind, 21, 25, 29, 138
gravity, force of, 21, 23, 24

halteres, 10, 140
hardening of the wing, 15
heat, 80, 81, 82, 106
helicopter, 47, 50, 92, 96, 128, 129,
  146, 149
Hierodula, 110, 118, 119
high speed camera, 30, 40, 45
hinge, basal, 10, 35, 61–69, 72, 104
Holst, Erich von, 145, 149
honey bee, 69, 72, 79, 88, 98, 120, 128,
  129, 132, 133, 140
honeysuckle, 13
hornet, 88, 120, 122
horsepower, 68, 69, 72, 77, 80, 88, 92,
  117
housefly, 9, 88, 110, 119, 120, 121, 146,
  147
hoverfly, 15, 72, 73, 92, 96
hummingbird, 13, 55, 58, 89, 90, 94,
  96, 128, 129
hummingbird hawk moth, 88, 92
hunting wasp, 57, 59, 60
Hyloicus, 15
hymenoptera, 13, 15, 93, 95, 102, 117,
  118

Icerya, 15
illustration, medieval, 115, 124, 127
indirect flight muscle, 61, 62, 65, 97,
  103, 104, 107, 108, 110, 115

innervation, multiple, 106, 108
insect, fastest, 85, 91
insect, smallest, 89, 93

jet aircraft, 28, 47, 69, 80, 91
Johnston's organ, 127, 141, 143

kinematics of wing beat, 35, 36, 39–47,
  49–52
kinetic energy, 82, 83, 88, 89
kymogram, 36, 37

landing, 110, 116, 119, 134, 136
Lathyrophthalmus, 15
leg, 38, 52, 55, 57, 58, 62, 72, 90, 91,
  116–120
longhorn beetle, 30, 31
Libellula quadrimaculata, 15, 58, 61
lift, 23–29, 36, 48, 49, 51, 52, 90, 97,
Lift balance, 35, 37, 114
lift, centre of, 23
lift, correlated with 108, 109
  stimulation frequency, 113, 114
lift, drag ratio, 28, 29
light, polarised, 133–135
light, ultraviolet, 133, 135
lime moth, 128
Limnophilus, 15
Lindauer, Martin, 142, 144
lines of weakness, 19, 20
  performed
Liothrips, 15
locust, 21, 28, 60, 61, 72, 77–80, 88,
  96–99, 103, 104, 108–110, 113, 122,
  124, 127–129, 139, 140

mandibles, 115
Mantoida, 15
mayfly, 55, 58, 59, 61, 62, 92, 95, 102,
  120, 122
mechanoreceptor, 132, 139
Megaloprepus coerulatus, 89
Melolontha, 127, 140, 141
microelectrode, 94, 101
micromanipulator, 94, 101
microscope, 72, 94, 101, 106, 107, 110,
  114
microswitch, 65–67, 69–71
microtome, 59, 65
midge, 140
  see gnat
migration, 79, 122, 124, 127–130
Miltogramma, 56, 60
mitochondria, 72, 74, 105–108, 111
monarch butterfly, 127, 128

morphology of insect, 9–11
morphology of wing, 13–20
mosquito, 88, 92, 116, 127–130
mud-dauber wasp, 15, 98, 99, 103, 110,
  111
muscle action potentials, 96, 97
  direct flight motor
muscle action potentials, 76, 77
  indirect flight motor, 97, 98, 105
muscle, power output, 68, 69, 72, 76,
  77, 78
muscle, ultrastructure, 72, 103, 105–
  108, 111, 112
Musculus latus, 68, 70, 71
myofibril, 107
myogenic rhythm, 103, 104
myosin, 106–109
Myrmica, 20

navigation, 126, 131–134
nectar, 130, 131
nervous system, 60–68, 93, 94, 96–
  centre of flight muscles, 107, 112,
  113
nervous system, 108–110, 113–114
  control of muscle power
nervous system, control of take-off,
  117, 118
nervous system, 139, 140
  information from sensory glands
neurogenic rhythm, 97, 103
Nielsen, E. T., 127

oleander moth, 128
olive fly, 142, 144
ommatidid, 133, 135
Oplomerus, 98, 99
Orneodes, 15
oxygen, 68, 69, 72, 77, 81, 83, 84, 105,
  107, 108

Panurgus calcaratus, 57, 60
Panorpa, 15
parascutum, 66, 67, 70, 71
Patasson crassicornis, 89, 93
peacock butterfly, 21, 55, 58
pedicel, 143
Perilampus, 15
petrol engine, 73, 80, 107, 117
Phasmidohelea, 20
Philanthus, 30, 35, 103, 108, 109
Phormia regina, 46, 49, 77, 82, 83
pigment cells, 133, 135
pleural plate (pleuron), 59, 66, 67
plume moth, 15, 62

point of action, 23, 24
polar diagram, 27–29
power, 69, 72, 77, 80, 88, 89, 91–94
praying mantis, 110, 118, 119
privet moth, 58, 92, 96
projection, 39, 40–42, 56
propeller, 20, 47, 48, 50, 145
pygmy insects, 89–91, 94

radar, 142, 144
Remer, Max, 133, 135
resilin, 82, 83, 88, 89, 146
rhabdome, 133, 135
Rhaphidia, 15
rhythm, myogenic, 103, 104
rhythm, neurogenic, 97
robberfly, 55, 58
rotational movements, 37, 45–49, 50–
     of wing, 52, 57, 115
ruby-throated hummingbird, 128, 129

sand wasp, 84, 86, 103, 108
scales, 19, 27
scape, 143
Schistocerca gregaria, 28, 96, 97, 102, 103,
     115, 124, 126, 127
Schremmer, Fritz, 57, 60
sclerite, 67, 70
scolopidia, 143
scorpion fly, 15

scutum, 66, 67, 70, 71, 117, 118
sensory organs, 139, 140
servomechanism, 37, 38, 137–139
side plate, see plural plate, 65, 70, 71
sidewind compensation, 139, 140
snake-fly, 15
sonar, 142–145
speed of flight, 37, 87, 88, 134, 136–139
Sphinx, 15
spurge hawk moth, 30, 35, 128
squama, 15
stag beetle, 22, 55, 58, 88, 92
starter muscle, 117
steering, see turning
Steiner, H., 78, 82
swarm, locust, 122–128, 140–143
swarm, midges, 141, 142
swift, 85, 91
synchronous flight muscles, see direct
     flight muscles
Syrphus, 72, 73

Taeniothrips, 10, 89, 94
take-off, 118, 140
tape-recorder, 94, 98, 109, 114
tarsul reflex, 38, 39
taxonomy, 12
thermal, 21, 22, 127
thrips, 35–38, 47–52, 56, 57
tiger beetle, 49, 122, 123
Tischner, Horst, 141

torsion, 45, 140
trachae, 65, 77, 105, 107
trehalose, 79, 83
T-system, 106, 108, 111
turning during flight, 77, 110

upcurrents, 21, 22
upstroke, 30, 42, 43, 45–52, 57, 64–67,
     70, 10
Urania, 10

veins, 15–17, 29
vinegar fly, 28, 29, 79, 90, 129, 143, 144

weakness, preformed line of, 19
weight, 23–25, 35, 49, 104
wind tunnel, 26, 35, 36, 38, 40, 41, 42,
     44, 46, 55
wing, artificial, 28, 29, 47, 57, 89, 91, 94
wing beat, frequency, 13, 60, 62, 65, 66,
     79, 91, 95–99, 103–4
wing beat, path of, see kinematics of
     wing beat
wing, bristly, 89–91
wing construction, 13–20, 62, 90
wing folding, 120–124
wing, polar diagram, 27–29, 48, 51

Zedwitz, Franz Graf, 58
Zorotypus, 10
Zschokke, Friedrich, 55, 58, 61